国家林业和草原局普通高等教育"十四五"规划教材

水土保持与荒漠化防治专业英语

马 瑞 主编

内 容 简 介

本书分2个部分：第1部分为Basic of Soil Erosion and Conservation（专业基础篇），共设计9个单元（Unit 1~9），分别为Unit 1 Outline of Soil Erosion（土壤侵蚀概述）、Unit 2 Water Erosion（水力侵蚀）、Unit 3 Wind Erosion（风力侵蚀）、Unit 4 Mass Erosion（重力侵蚀）、Unit 5 Sandy Desertification（荒漠化）、Unit 6 Rocky Desertification（石漠化）、Unit 7 Dust Storm（沙尘暴）、Unit 8 Ecological Engineering（生态工程）、Unit 9 Watershed Management（流域管理）；第2部分为Academic English Writing（学术英语写作），共2个单元（Unit 10~11），分别为Unit10 Characters of Academic English（学术英语特点）和Unit 11 Writing up Research（研究报告撰写）。

本书主要作为水土保持与荒漠化防治一级学科下设本科专业的"专业英语"课程教材，也可作为水利、林学、自然地理、环境等学科专业研究生教学用书。

图书在版编目(CIP)数据

水土保持与荒漠化防治专业英语 / 马瑞主编.
北京：中国林业出版社，2024.12. — （国家林业和草原局普通高等教育"十四五"规划教材）. — ISBN 978-7-5219-2966-9

Ⅰ.S157；P942.073

中国国家版本馆CIP数据核字第2024W5V136号

责任编辑：范立鹏
责任校对：苏　梅
封面设计：周周设计局

出版发行：中国林业出版社
　　　　　（100009，北京市西城区刘海胡同7号，电话83143626）
电子邮箱：jiaocaipublic@163.com
网址：https://www.cfph.net
印刷：北京中科印刷有限公司
版次：2024年12月第1版
印次：2024年12月第1次
开本：787mm×1092mm　1/16
印张：10.375
字数：250千字
定价：38.00元

《水土保持与荒漠化防治专业英语》编写人员

主　　编　马　瑞
副 主 编　贾玉华　鱼燕萍　马迎梅
编写人员　（按姓氏笔画排序）

　　　　　　马　瑞（甘肃农业大学）

　　　　　　马迎梅（内蒙古农业大学）

　　　　　　牛耀彬（山西农业大学）

　　　　　　孙　迪（沈阳农业大学）

　　　　　　鱼燕萍（甘肃农业大学）

　　　　　　赵龙山（贵州大学）

　　　　　　贾玉华（沈阳农业大学）

　　　　　　梁心蓝（四川农业大学）

　　　　　　韩　珍（贵州大学）

　　　　　　解婷婷（甘肃农业大学）

前　言

2018年，本教材立项为甘肃农业大学教材建设项目，并于2020年获批列入国家林业和草原局普通高等教育"十四五"规划教材。为保质保量完成教材编写任务，我们组织甘肃农业大学、沈阳农业大学、内蒙古农业大学、山西农业大学、贵州大学和四川农业大学6所高校共10位专业教师组成了本教材编写组。

编者收集整理了维基百科(Wikipedia)、专著、文献中关于水土保持与荒漠化防治领域的相关资料，形成了由专业基础篇和学术英语写作两部分内容构成的教材。本教材编写力求内容全面、章节合理。专业基础篇以专业论著的读写能力提升为出发点，介绍了主要土壤侵蚀类型、环境问题及生态修复内容，旨在帮助学生提高专业英语文献的阅读理解能力，提升翻译技巧及写作技能，同时巩固和拓展学生的专业知识；学术英语写作主要针对SCI论文撰写，内容包括学术英语特点和研究报告撰写两部分，旨在提高学生对学术论文基本框架的认识、提升内容组织能力和写作技巧；附录汇总了水土保持与荒漠化防治领域高频专业词汇，旨在使学生在听说读写过程中正确理解、翻译和使用专业词汇。

本教材编写分工如下：Unit 1由马瑞编写；Unit 2由梁心蓝和赵龙山编写；Unit 3由鱼燕萍和马迎梅编写；Unit 4由贾玉华编写；Unit 5由牛耀彬编写；Unit 6由韩珍和赵龙山编写；Unit 7由孙迪编写；Unit 8和Unit 9由贾玉华编写；Unit 10由解婷婷编写；Unit 11由马迎梅和鱼燕萍编写；Appendix由鱼燕萍编写。全书最后由马瑞统稿和定稿。

限于编写时间，本教材难免留有不妥或错漏之处，恳请广大读者批评指正。

马　瑞
2024年10月于兰州

CONTENTS

Chapter 1 Basic of Soil Erosion and Conservation

Unit 1 Outline of Soil Erosion (2)
- 1.1 Conception (2)
- 1.2 Forms & features of main soil erosion types (2)
  - 1.2.1 Water erosion (2)
  - 1.2.2 Wind erosion (3)
  - 1.2.3 Mass erosion (3)
- 1.3 Factors affecting soil erosion (4)
  - 1.3.1 Natural factors (4)
  - 1.3.2 Human activities increase soil erosion (5)
  - 1.3.3 Climate change changes soil erosion intensity (6)
- 1.4 Site effects of soil erosion (7)
- 1.5 Global environmental effects (8)
  - 1.5.1 Land degradation (8)
  - 1.5.2 Sedimentation of aquatic ecosystems (9)
  - 1.5.3 Airborne dust pollution (9)
- 1.6 Monitoring, measuring and modeling soil erosion (9)
- 1.7 Prevention and remediation (10)

Unit 2 Water Erosion (12)
- 2.1 Rill erosion (13)
  - 2.1.1 Conception (13)
  - 2.1.2 Rills created by erosion (13)
  - 2.1.3 Rill initiation (13)
  - 2.1.4 Significance of rill erosion (14)
- 2.2 Gully erosion (15)
  - 2.2.1 Conception (15)

CONTENTS

    2.2.2 Formation and consequences ·· (16)
  2.3 Headward erosion ·· (17)
    2.3.1 Conception ·· (17)
    2.3.2 Stream types created by headward erosion ·· (18)
    2.3.3 Drainage patterns created by headward erosion ·· (18)
  2.4 Debris flow ·· (19)
    2.4.1 Conception ·· (19)
    2.4.2 Features and behavior ·· (19)

Unit 3 Wind Erosion ·· (23)
  3.1 Conception ·· (23)
  3.2 Transport ·· (24)
  3.3 Deposition ·· (25)
  3.4 Aeolian landform ·· (26)
    3.4.1 The types of wind erosion landforms ·· (26)
    3.4.2 The types of wind deposition landforms ·· (30)

Unit 4 Mass Erosion ·· (39)
  4.1 Overview ·· (39)
  4.2 Sinkholes ·· (39)
  4.3 Earthflows ·· (40)
  4.4 Landslides ·· (41)

Unit 5 Sandy Desertification ·· (45)
  5.1 Conception ·· (45)
  5.2 Areas affected ·· (45)
  5.3 Causes and consequences ·· (46)
    5.3.1 Vegetation destruction ·· (46)
    5.3.2 Climate change ·· (46)
    5.3.3 Overgrazing ·· (47)
    5.3.4 Irrigated croplands ·· (47)
  5.4 Countermeasures and prevention ·· (48)
    5.4.1 Reforestation ·· (48)
    5.4.2 Enriching of the soil and restoration fertility ·· (49)
    5.4.3 Farmer-managed natural regeneration ·· (49)
    5.4.4 Carbon trading and carbon sequestration ·· (49)
    5.4.5 Sustainable land and soil management ·· (50)

Unit 6　Rocky Desertification …… (53)
6.1　Conception …… (53)
6.2　Causes …… (53)
6.2.1　Natural processes …… (54)
6.2.2　Human factor …… (55)
6.3　Distributions …… (56)
6.4　Environmental, social and economic hazard …… (56)
6.4.1　Environmental hazards …… (57)
6.4.2　Social hazards …… (57)
6.4.3　Economic hazards …… (58)

Unit 7　Dust Storm …… (61)
7.1　Conception …… (61)
7.2　Causes …… (61)
7.3　Environmental and physical effects …… (62)
7.4　Source areas …… (63)
7.5　Global health impacts of dust storms …… (65)
7.5.1　Short-term health effects …… (65)
7.5.2　Long-term health effects …… (68)
7.6　Control of dust storms …… (68)

Unit 8　Ecological Engineering …… (71)
8.1　Overview …… (71)
8.2　Terrace …… (72)
8.3　Conservation tillage …… (74)
8.4　Windbreak …… (75)
8.5　Sand fence …… (76)

Unit 9　Watershed Management …… (81)
9.1　Overview …… (81)
9.2　Integrated water resources management …… (82)
9.3　Environmental laws …… (83)

Chapter 2　Academic English Writing

Unit 10　Characters of Academic English …… (90)
10.1　Vocabulary characteristics …… (90)
10.2　Features of grammatical structure …… (90)

10.3	Features of commonly used sentence	(92)
10.4	Exercises	(94)

Unit 11 Writing up Research .... (100)

11.1	Formulating a research question	(101)
11.2	Title	(101)
11.3	Abstract	(102)
11.4	Introduction	(107)
11.5	Materials and methods	(112)
	11.5.1 Materials	(112)
	11.5.2 Methods	(114)
11.6	Results	(116)
11.7	Discussion	(117)
11.8	Acknowledgments	(119)
11.9	References	(119)

References .... (120)

Appendix .... (122)

Chapter 1

Basic of Soil Erosion and Conservation

Unit 1　Outline of Soil Erosion

1.1　Conception

Soil erosion is the displacement of the upper layer of soil. It is one form of soil degradation. It is a process consisting of the detachment of individual soil particles from the soil mass and their transport by erosive agents such as water and wind. When suffificient energy of erosive agents is no longer available to transport the particles, deposition occurs. Soil erosion may be a slow process that continues relatively unnoticed, or it may occur at an alarming rate causing a serious loss of topsoil.

These agents that cause soil erosion include water, wind, gravity, temperature, glaciers, chemical reaction, plants, animals, and humans. In accordance with the erosive agents, erosion is divided into water erosion, wind (aeolian) erosion, mass erosion, freezing-thawing erosion, glacial erosion, chemical erosion, bioerosion, anthropogenic erosion and mixed erosion.

1.2　Forms & features of main soil erosion types

1.2.1　Water erosion

(1) Rainfall and surface runoff erosion

Rainfall, and the surface runoff which may result from rainfall, produces four main types of soil erosion: splash erosion, sheet erosion, rill erosion, and gully erosion. Splash erosion is generally seen as the first and least severe stage in the soil erosion process, which is followed by sheet erosion, then rill erosion and finally gully erosion.

Rainsplash is the most important detaching agent. In splash erosion, the impact of a falling raindrop creates a small crater in the soil, ejecting soil particles. The distance these soil particles travel can be as much as 0.6 m (two feet) vertically and 1.5 m horizontally on level ground. If the soil is saturated, or if the rainfall rate is greater than the rate at which water can infiltrate into the soil, surface runoff occurs. If the runoff has sufficient flow energy, it will transport loosened soil particles down the slope. Sheet erosion is the transport of loosened soil particles by overland flow. Rill erosion refers to the development of small, ephemeral concentrated flow paths which function as both sediment source and sediment delivery systems for erosion on hillslopes. Gully erosion occurs when runoff water accumulates and rapidly flows in narrow channels during or immediately after heavy rains or melting snow, removing soil to a considerable depth.

(2) Streams erosion

Stream erosion occurs with continued water flow. The erosion is both downward, deepening the valley, and headward, extending the valley into the hillside, creating head cuts and steep banks. In the earliest stage of stream erosion, the erosive activity is dominantly vertical erosion, the valleys have a typical V cross-section and the stream gradient is relatively steep. When some erosion base level is reached, the erosive activity switches to lateral erosion, which widens the valley floor. Then, the valley bottom gradient, becomes nearly flat, and lateral deposition of sediments becomes important as the stream meanders across the valley floor.

In all stages of stream erosion, the most erosion occurs during times of flood, when more and faster-moving water is available to carry a larger sediment. In such processes, it is not the water alone that erodes, suspended abrasive particles, pebbles and boulders can also act erosively as they traverse a surface.

1.2.2 Wind erosion

Wind erosion is a major geomorphological force, especially in arid and semi-arid regions. It is also a major cause of land degradation, desertification, harmful airborne dust, dust storms and crop damage, especially after human activities such as deforestation, overgrazing, agricultural practices and urbanization have intensified it far beyond natural levels.

Wind erosion is of two primary varieties: deflation, where the wind picks up and carries away loose particles; and abrasion, where surfaces are worn down as they are struck by airborne particles carried by wind. Deflation is divided into three categories: surface creep, where larger, heavier particles slide or roll along the ground; saltation, where particles are lifted a short height into the air, and bounce and saltate across the surface of the soil; and suspension, where very small and light particles are lifted into the air by the wind, and are often carried for long distances. Saltation is responsible for the majority (50%~70%) of wind erosion, followed by suspension (30%~40%), and then surface creep (5%~25%).

1.2.3 Mass erosion

Mass erosion, also called gravity erosinon, is the downward and outward movement of rock and sediments on a sloped surface due to the force of gravity.

Mass movement is often the first stage in the breakdown and transport of weathered materials in mountainous areas. It moves material from higher elevations to lower elevations where other eroding agents such as streams and glaciers can then pick up the material and move it to even lower elevations. Mass movement processes are always occurring continuously on all slopes; some mass movement processes act very slowly; others occur very suddenly, often with disastrous results. In general terms, any perceptible downslope movement of rock or sediment is often referred to as a landslide. Surface creep is the slow movement of soil and rock debris by gravity which is usually not perceptible except through long term observation. However, landslides often occur in an instant. Landslide happens on steep hillsides, occurring along distinct fracture zones where materials, once released, may move quite rapidly downhill.

1.3 Factors affecting soil erosion

1.3.1 Natural factors

(1) Climate

For water erosion, the intensity and amount of precipitation is the main climatic factor governing erosion. The relationship is particularly strong if heavy rainfall occurs at times when, or in locations where, the soil's surface is not well protected by vegetation. For instance, during periods when agricultural activities leave the soil bare, or in arid and semi-arid regions where vegetation is naturally sparse. In some areas, rainfall intensity is the primary determinant of erosivity, with higher intensity rainfall generally resulting in more soil erosion by water. The size and velocity of rain drops is also an important factor. Larger and higher-velocity rain drops have greater kinetic energy, and thus their impact will displace soil particles by larger distances than smaller, slower-moving rain drops. In other areas, runoff and resulting erosion result from relatively low intensities of rainfall falling onto previously saturated soil. In such situations, rainfall amount rather than intensity is the main factor determining the severity of soil erosion by water.

For wind erosion, strong winds is the main climatic factor governing erosion, particularly during times of drought when vegetation is sparse and soil is dry. Other climatic factors such as average temperature and temperature range may also affect erosion, via their effects on vegetation and soil properties. In general, given similar vegetation, areas with more precipitation, especially high-intensity rainfall, or areas with more strong wind are expected to have more soil erosion.

(2) Soil structure and composition

The composition, moisture, and compaction of soil are all major factors in determining the erosivity of rainfall. Because the clay helps bind soil particles together, sediments containing more clay tend to be more resistant to erosion than those with sand or silt. Because the organic materials coagulate soil colloids and create a stronger, more stable soil structure, soil containing high levels of organic materials are often more resistant to erosion. The amount of water present in the soil before the precipitation also plays an important role, because it sets limits on the amount of water that can be absorbed by the soil. Wet, saturated soils will not be able to absorb as much rain water, leading to higher levels of surface runoff and thus higher erosivity for a given volume of rainfall. Soil compaction also affects the permeability of the soil to water, and hence the amount of water that flows away as runoff. More compacted soils will have a larger amount of surface runoff than less compacted soils.

(3) Vegetative cover

Vegetation acts as an interface between the atmosphere and the soil. It increases the permeability of the soil to rainwater, thus decreasing runoff. It shelters the soil from winds, which results in decreased wind erosion, as well as advantageous changes in microclimate. The roots of the plants bind the soil together, and interweave with other roots, forming a more solid mass that is less susceptible to both water and wind erosion. The removal of vegetation increases the rate of surface erosion.

(4) Topography

The topography of the land determines the velocity at which surface runoff will flow, which in turn determines the erosivity of the runoff. Longer, steeper slopes, especially those without adequate vegetative cover, are more susceptible to very high rates of erosion during heavy rains than shorter, less steep slopes. Steeper terrain is also more prone to mudslides, landslides, and other forms of gravitational erosion processes.

1.3.2 Human activities increase soil erosion

(1) Agricultural practices

Unsustainable agricultural practices are the single greatest contributor to the global increase in erosion rates. The tillage of agricultural lands, which breaks up soil into finer particles, is one of the primary factors. The problem has been exacerbated in modern times, due to mechanized agricultural equipment that allows for deep plowing, which severely increases the amount of soil that is available for transport by water erosion.

Tillage also increases wind erosion rates, by dehydrating the soil and breaking it up into smaller particles that can be picked up by the wind. Exacerbating this is the fact that most of the trees are generally removed from agricultural fields, allowing winds to have long, open runs to travel over at higher speeds.

Other agricultural practices, including monocropping, farming on steep slopes, pesticide and chemical fertilizer usage, and the use of surface irrigation, are all responsible for the soil erosion. For the grassland, heavy grazing reduces vegetative cover as well as causes severe soil compaction, both of which increase erosion rates.

(2) Deforestation

In an undisturbed forest, the mineral soil is protected by a layer of leaf litter and an humus that cover the forest floor. These two layers form a protective mat over the soil that absorbs the kinetic energy of rain drops. They are porous and highly permeable to rainfall, and allow rainwater to slow percolate into the soil below, instead of flowing over the surface as runoff. The roots of the trees and plants hold together soil particles, preventing them from being washed away. The vegetative cover acts to reduce the velocity of the raindrops that strike the foliage and stems before hitting the ground, reducing their kinetic energy. However, it is the forest floor, more than the canopy, that prevents surface erosion. The terminal velocity of rain drops is reached in about 8 m. Because forest canopies are usually higher than this, rain drops can often regain terminal velocity even after striking the canopy. However, the intact forest floor, with its layers of leaf litter and organic matter, is still able to absorb the impact of the rainfall.

Deforestation causes increased erosion rates due to exposure of mineral soil by removing the humus and litter layers from the soil surface, removing the vegetative cover and plant roots that binds soil together, and causing heavy soil compaction from logging equipment. Once trees have been removed by logging, infiltration rates become low and erosion rates become high.

(3) Roads and urbanization

Urbanization has major effects on erosion processes: first by denuding the land of vegetative cover, altering drainage patterns, and compacting the soil during construction; and next by covering the land in an impermeable layer of asphalt or concrete that increases the amount of surface runoff and increases surface wind speeds. Much of the sediment carried in runoff from urban areas, especially from roads, is highly contaminated with fuel, oil, and other chemicals. This increased runoff, in addition to eroding and degrading the land that it flows over, also causes major disruption to surrounding watersheds by altering the volume and rate of water that flows through them, and filling them with chemically polluted sedimentation. The increased flow of water through local waterways also causes a large increase in the rate of bank erosion.

1.3.3 Climate change changes soil erosion intensity

The warmer atmospheric temperatures observed over the past decades are expected to lead to a more vigorous hydrological cycle, including more extreme rainfall events. If rainfall amounts and intensities increase in many parts of the world as expected, erosion will also increase, unless amelioration measures are taken. Studies indicated that, other factors such as land use unconsidered, it is reasonable to expect approximately a 1.7% change in soil erosion for each 1% change in total precipitation under climate change.

Soil erosion rates are expected to change in response to climate change for a variety of reasons. These reasons include: ①The erosive power of rainfall increase; ②Changes in plant canopy caused by plant biomass production associated with moisture regime; ③Changes in litter cover on the ground caused by both plant residue decomposition rates driven by microbial activity related to soil temperature and moisture as well as plant biomass production rates; ④Changes in soil moisture due to precipitation regimes and evapotranspiration rates, which changes infiltration and runoff ratios; ⑤Soil erodibility increase due to decrease in soil organic matter concentrations in soils that lead to a more susceptible soil structure to erosion. Besides, runoff also increase due to increased soil surface sealing and crusting; ⑥A shift of winter precipitation from non-erosive snow to erosive rainfall due to increasing winter temperatures; ⑦Melting of permafrost, which turn a previously non-erodible soil state to an erodible one.

1.4 Site effects of soil erosion

Soil erosion has both on-site and off-site effects.

On-site effects include decreases in agricultural productivity and ecological collapse, both because of loss of the nutrient-rich upper soil layers. In some cases, the eventual end result is desertification. on-site effects are particularly important on agricultural land where the redistribution of soil within a field, the loss of soil from a field, the breakdown of soil structure and the decline in organic matter and nutrient result in a reduction of cultivable soil depth and a decline in soil fertility. Erosion also reduces available soil moisture, resulting in more drought-prone conditions. The net effect is a loss of productivity, which restricts what can be grown and results in increased expenditure on fertilizers to maintain yields. If fertilizers were used to compensate for loss of fertility arising from erosion, there would be a substantial hidden cost to the farmers. The loss of soil fertility through erosion ultimately leads to the abandonment of land, with consequences for food production and food security and a substantial decline in land value.

Off-site effects include sedimentation of waterways and eutrophication of water bodies, as well as sediment-related damage to roads and houses, et al. Off-site problems arise from sedimentation downstream or downwind, which reduces the capacity of rivers and drainage ditches, enhances the risk of flooding, blocks irrigation canals and shortens the design life of reservoirs. Many irrigation projects have been ruined as a consequence of erosion. Sediment is also a pollutant in its own right and, through the chemicals adsorbed to it, can increase the levels of nitrogen and phosphorus in water bodies and result in eutrophication. Erosion leads to the breakdown of soil aggregates and clods into their primary particles of clay, silt and sand. Through this process, the carbon that is held within the clays and the soil organic content is released into the atmosphere as CO_2. Erosion is therefore a contributor to climatic change, since increasing the carbon dioxide content of the atmosphere enhances the greenhouse effect.

Water and wind erosion are the two primary causes of land degradation; combined, they are responsible for about 84% of the global extent of degraded land, making excessive erosion (abnormal erosion) one of the most significant environmental problems worldwide.

1.5 Global environmental effects

Due to the severity of its ecological effects, and the scale on which it is occurring, soil erosion constitutes one of the most significant global environmental problems we face today.

1.5.1 Land degradation

Water and wind erosion are now the two primary causes of land degradation; combined, they are responsible for 84% of degraded acreage.

Each year, about 7500×10^8 t of soil is eroded from the land—a rate that is about 13~40 times as fast as the natural rate of erosion. Approximately 40% of the world's agricultural land is seriously degraded. According to the United Nations, an area of fertile soil the size of Ukraine is lost every year because of drought, deforestation and climate change. In Africa, if current trends of soil degradation continue, the continent might be able to feed just 25% of its population by 2025, according to UNU's Ghana-based Institute for Natural Resources in Africa.

Recent modeling developments have quantified rainfall erosivity at global scale using high temporal resolution (<30 min) and high fidelity rainfall recordings. The result is an extensive global data collection effort produced the Global Rainfall Erosivity Database (GloREDa) which includes rainfall erosivity for 3625 stations and covers 63 countries. This first ever Global Rainfall Erosivity Database was used to develop a global erosivity map at 30 arc-seconds (1 km) based on sophisticated geostatistical process. According to a new study published in Nature Communications, almost 3600×10^4 of soil is lost every year due to water, and deforestation and other changes in land use make the problem worse. The study investigates global soil erosion dynamics by means of high-resolution spatially distributed modelling (ca. 250 m×250 m cell size). The geo-statistical approach allows, for the first time, the thorough incorporation into a global soil erosion model of land use and changes in land use, the extent, types, spatial distribution of global croplands and the effects of different regional cropping systems.

The loss of soil fertility due to erosion is further problematic because the response is often to apply chemical fertilizers, which leads to further water and soil pollution, rather than to allow the land to regenerate.

1.5.2 Sedimentation of aquatic ecosystems

Soil erosion, especially from agricultural activity, is considered to be the leading global cause of diffuse water pollution, due to the effects of the excess sediments flowing into the world's waterways. The sediments themselves act as pollutants, as well as being carriers for other pollutants, such as attached pesticide molecules or heavy metals.

The effect of increased sediments loads on aquatic ecosystems can be catastrophic. Silt can smother the spawning beds of fish, by filling in the space between gravel on the stream bed. It also reduces their food supply, and causes major respiratory issues for them as sediment enters their gills. The biodiversity of aquatic plant and algal life is reduced, and invertebrates are also unable to survive and reproduce. While the sedimentation event itself might be relatively short-lived, the ecological disruption caused by the mass die off often persists long into the future.

One of the most serious and long-running water erosion problems worldwide is in the People's Republic of China, on the middle reaches of the Yellow River and the upper reaches of the Yangtze River. From the Yellow River, a mass of sediment flows into the ocean each year. The sediment originates primarily from water erosion in the Loess Plateau region of the northwest.

1.5.3 Airborne dust pollution

Soil particles picked up during wind erosion are a major source of air pollution, in the form of airborne particulates—dust. These airborne soil particles are often contaminated with toxic chemicals such as pesticides or petroleum fuels, posing ecological and public health hazards when they later land, or are inhaled/ingested.

1.6 Monitoring, measuring and modeling soil erosion

Monitoring and modeling of erosion processes can help people better understand the causes of soil erosion, make predictions of erosion under a range of possible conditions, and plan the implementation of preventative and restorative strategies for erosion. However, the complexity of erosion processes and the number of scientific disciplines that must be considered to understand and model them (e. g. climatology, hydrology, geology, soil science, agriculture, chemistry, physics, etc.) makes accurate modelling challenging. Erosion models are also non-linear, which makes them difficult to work with numerically, and makes it difficult or impossible to scale up to making predictions about large areas from data collected by sampling smaller plots.

The most commonly used model for predicting soil loss from water erosion is the Universal Soil Loss Equation (USLE). This was developed in the 1960s and 1970s. It estimates the average annual soil loss A on a plot-sized area as:

$$A = RKLSCP$$

where R is the rainfall erosivity factor; K is the soil erodibility factor; L and S are topographic factors representing length and slope; C is the cover and management factor and P is the support practices factor.

Despite the USLE's plot-scale spatial focus, the model has often been used to estimate soil erosion on much larger areas, such as watersheds or even whole continents. One major problem is that the USLE cannot simulate gully erosion, and so erosion from gullies is ignored in any USLE-based assessment of erosion. Yet erosion from gullies can be a substantial proportion (10%~80%) of total erosion on cultivated and grazed land.

During the 50 years since the introduction of the USLE, many other soil erosion models have been developed. But because of the complexity of soil erosion and its constituent processes, all erosion models fail to give satisfactory results when validated i.e. when model predictions are compared with real-world measurements of erosion. Thus new soil erosion models continue to be developed. Some of these remain USLE-based. Other soil erosion models have largely or wholly abandoned usage of USLE elements.

1.7 Prevention and remediation

The most effective known method for erosion prevention is to increase vegetative cover on the land, which helps prevent both wind and water erosion. Terracing is an extremely effective means of erosion control, which has been practiced for thousands of years by people all over the world. Windbreaks, also called shelterbelts, are rows of trees and shrubs that are planted along the edges of agricultural fields, to shield the fields against winds. In addition to significantly reducing wind erosion, windbreaks provide many other benefits such as improved microclimates for crops which are sheltered from the dehydrating and otherwise damaging effects of wind, habitat for beneficial bird species, carbon sequestration, and aesthetic improvements to the agricultural landscape. Traditional planting methods, such as mixed cropping and crop rotation have also been shown to significantly reduce erosion rates. Crop residues play a role in the mitigation of erosion, because they reduce the impact of raindrops breaking up the soil particles. There is a higher potential for erosion when producing potatoes than when growing cereals, or oilseed crops. Forages have a fibrous root system, which helps combat erosion by anchoring the plants to the top layer of the soil, and covering the entirety of the field, as it is a non-row crop. In tropical coastal systems, properties of mangroves have been examined as a potential means to reduce soil erosion. Their complex root structures are known to help reduce wave damage from storms and flood impacts while binding and building soils. These roots can slow down water flow, leading to the deposition of sediments and reduced erosion rates. However, in order to maintain sediment balance, adequate mangrove forest width needs to be present.

Extending reading: Factors influencing soil erosion

On the world scale, investigations of the relationship between soil loss and climate show that at annual precipitation totals below 450 mm, erosion increases as precipitation increases. But as precipitation increases so does the vegetation cover, resulting in better protection of the soil surface, so that for annual precipitation between 450 mm and 650 mm, soil loss decreases as precipitation increases. However, further increases in precipitation are sufficient to overcome the protective effect and erosion then increases until, again, the vegetation responds by becoming sufficiently dense to provide additional protection, causing erosion to decrease. Above 1700 mm, the volume and intensity of the rain outweigh the protective effect of the vegetation and erosion increases with precipitation.

On the local scale, The factors controlling soil erosion are the erosivity of the eroding agent, the erodibility of the soil, the slope of the land and the nature of the plant cover. The factors may be considered under three headings: energy, resistance and protection. The energy group includes the potential ability of rainfall, runoff and wind to cause erosion. This ability is termed erosivity. Also included are those factors that directly affect the power of the erosive agents, such as the reduction in the length of runoff or wind blow through the construction of terraces and wind breaks respectively. Fundamental to the resistance group is the erodibility of the soil, which depends upon its mechanical and chemical properties. Factors that encourage the infiltration of water into the soil and thereby reduce runoff decrease erodibility, while any activity that pulverizes the soil increase it. Thus cultivation may decrease the erodibility of clay soils but increase that of sandy soils. The protection group focuses on factors relating to the plant cover. By intercepting rainfall and reducing the velocity of runoff and wind, plant cover can protect the soil from erosion. Different plant cover affords different degrees of protection, so, by determining land use, the human can control the rate of erosion to a considerable degree.

Source:
Morgan R P C. Soil Erosion & Conservation, 3rd ed. 2005.

Unit 2 Water Erosion

Water erosion is the process by which soil is removed from the earth's surface through the action of water. This type of erosion occurs primarily due to rainfall and surface runoff. It begins with the impact of raindrops on the soil, which disrupts soil particles, a process known as splash erosion. Once detached, these particles can be transported by flowing water.

There are several forms of water erosion(Fig. 2-1), including:

Splash erosion: Where raindrop impact dislodges and displaces soil particles on the surface, initiating soil erosion.

Sheet erosion: This occurs when a thin layer of soil is removed over a large area by surface runoff.

Rill erosion: Occurring in small channels on a slope, rill erosion forms when runoff water carves noticeable paths into the soil.

Gully erosion: When rills grow larger, they develop into gullies, which are deeper channels and can remove significant amounts of soil.

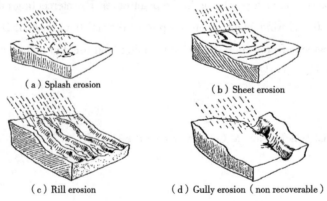

(a) Splash erosion (b) Sheet erosion
(c) Rill erosion (d) Gully erosion (non recoverable)

Fig. 2-1 Soil erosion types

2.1 Rill erosion

2.1.1 Conception

Rill erosion (Fig. 2-2) is a form of soil erosion that commonly occurs on sloping farmland. It involves the transformation of sheet flow from rainfall runoff into concentrated rill flow, which erodes the soil as it moves downhill. These small rills can often be filled by normal farming activities. In hillslope geomorphology, a rill is defined as a shallow channel, no more than a few tens of centimeters deep, formed by the erosive action of flowing water. Smaller incised channels are known as microrills, while larger channels are referred to as gullies.

Fig. 2-2 Rill erosion

2.1.2 Rills created by erosion

Rills are narrow and shallow channels eroded into unprotected soil by hillslope runoff. As rain continues, surface water gathers to form streams concentrated in grooves, increasing their erosive energy. This process not only washes away the soil but also causes lateral erosion, continuously altering the shape of the grooves and forming varied erosion patterns. Because soil is often left bare during agricultural operations, rills can form during these vulnerable periods. Moreover, rills may develop in areas where soil is exposed after deforestation or during construction activities.

Rills are easily visible when formed and serve as an early indicator of ongoing erosion. Without soil conservation measures, rills in frequently eroding areas can develop into larger features such as gullies or, in semi-arid regions, badlands.

2.1.3 Rill initiation

Rills are formed when water erodes topsoil on hillsides and are significantly influenced by seasonal weather patterns, appearing more frequently in wetter months. Rills begin to form when runoff shear stress—the force of surface runoff sufficient to detach soil particles—exceeds the soil's shear strength, which is the soil's resistance to forces parallel to its surface. This initiates the erosion process as water detaches and carries soil particles downslope. This susceptibility is more pronounced in sandy, loamy soils compared to denser clays, which resist rill formation.

Rill formation is also closely linked to the steepness of hillside slopes, as gravity provides the force needed to start the erosion. Therefore, the primary control on rill formation is slope steepness, which affects rill depth; the slope length and soil permeability influence the number of incisions. Every soil type has a threshold slope angle where water velocity is insufficient to create rills, often around 2 degrees with a shear velocity between 3 and 3.5 cm/s on non-cohesive slopes.

Once rills form, they are affected by additional forces that may increase their size and sediment output. Up to 37% of erosion in rill-dense areas may be due to mass movements or collapses of rill sidewalls, which occur when flowing water undercuts the walls or when water infiltration weakens them, leading to collapse.

Fig. 2-3 shows the four stage of erosion rill profile:

①Water erosion ditch stage: In this stage, water flow begins to erode the surface, forming shallow grooves or ditches that can rapidly expand after rainfall.

②Cutting stage of gully top: Erosion intensifies in this stage, leading to significant downcutting at the gully top. The grooves become more pronounced and deeper, further reducing the stability of the surface soil.

③Making balanced profile stage: Over time, the gully's profile reaches a more balanced state. The rates of erosion and sedimentation achieve a dynamic equilibrium, and changes in the gully shape become more gradual.

④Stop stage: Eventually, the erosion process slows down or ceases, resulting in minimal further development of the gully. At this point, the restoration of vegetation and human interventions may aid in stabilizing the terrain.

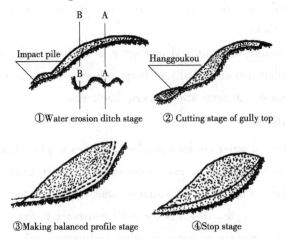

Fig. 2-3 Development stage of erosion rill profile

2.1.4 Significance of rill erosion

Rill erosion is a common and significant form of soil erosion that has profound effects on agricultural production, environmental protection, and land management. It typically occurs during surface runoff caused by rainfall or irrigation, where water flows along slightly depressed channels, concentrating erosion forces in narrow areas and thus intensifying soil loss. Compared to sheet erosion, rill erosion is more destructive because its effect is more concentrated and has greater depth, often resulting in significant changes to the landscape.

The impact of rill erosion on agriculture is considerable. It can quickly remove the fertile topsoil, leading to reduced soil fertility, thinning of the tillage layer, and consequently affecting crop growth and yields. The presence of rills may also disrupt the root distribution of crops, preventing plants from fully absorbing water and nutrients. Furthermore, the elongation and deepening of rills can disrupt the evenness of farmland, making cultivation more difficult and decreasing overall farm productivity.

From an environmental perspective, rill erosion adversely affects water quality by increasing sediment loads in rivers and lakes. These sediments can carry fertilizers, pesticides, and other pollutants from farmland, leading to eutrophication and impacting the health of aquatic ecosystems. Additionally, large amounts of sediment deposition can reduce the navigability and storage capacity of water bodies, increasing the risk of flooding.

In terms of land management, controlling and preventing rill erosion is crucial for achieving sustainable land use. Implementing suitable land management practices—such as contour farming, vegetation cover, building retention basins, and strategically planning water flow paths—can effectively reduce the formation and development of rills. Moreover, monitoring and assessing erosion activities to address severely affected areas can significantly improve land health. Therefore, the study and management of rill erosion are vital not only for maintaining agricultural productivity and ecological balance but also for the protection of soil resources and sustainable development.

2.2 Gully erosion

2.2.1 Conception

A gully is a landform created by running water, which sharply erodes soil, typically on hillsides. Gullies resemble large ditches or small valleys but aremeters to tens of meters in depth and width. During gully formation, substantial water flows cause significant deep cutting into the soil.

Gullies are relatively permanent, steep-sided watercourses experiencing ephemeral flows during rainstorms. Compared to stable river channels, which have a smooth, concave-long profile, gullies feature a headcut and various steps or knick-points. These rapid slope changes alternate with gentle gradient sections, either straight or slightly convex in profile. Gullies have relatively greater depth and smaller width than stable channels, carry larger sediment loads, and exhibit erratic behavior, often leading to poor relationships between sediment discharge and runoff. A commonly used distinction between gullies and rills is a cross-sectional area greater than 1 m^2 (929 cm^2) (Fig. 2-4), with gullies usually indicating accelerated erosion and landscape instability.

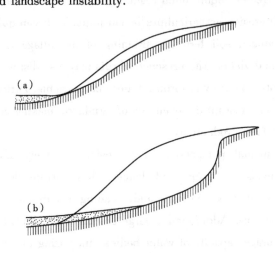

Fig. 2-4 Rill (a) and gully (b)

2.2.2 Formation and consequences

Gully erosion involves the formation of gullies. Hillsides become more prone to gully erosion when cleared of vegetation due to deforestation, over-grazing, or other means, with eroded soil easily carried by water during intense storms like thunderstorms.

Previously thought to develop from enlarged rills, studies in the Southwest USA suggest gully initiation is more complex. Initially, small depressions or knicks form on hillsides due to vegetation weakening from grazing or fire. Water concentrates in these depressions, enlarging them until multiple depressions coalesce to form an initial channel. Erosion concentrates at the depression heads, where near-vertical scarps develop, with most erosion occurring through scouring at the scarp base, leading to channel deepening and headwall collapse. Sediment is produced further down by bank erosion, through scouring water action or bank collapse. Between flows, sediment becomes erodible from weathering and bank collapse. This gully formation sequence, described by Leopold et al. (1964) in New Mexico, illustrates gully headward growth through interflow and surface runoff, as shown in Fig. 2-5.

Gullies reduce farmland productivity by incising the land and producing sediment that may clog downstream water bodies. Therefore, much effort is invested in studying gullies within geomorphology, preventing gully erosion, and restoring gullied landscapes. The total soil loss from gully formation and resulting downstream sedimentation can be significant.

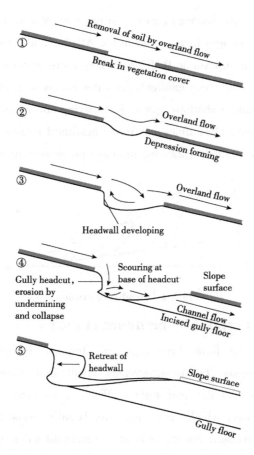

Fig. 2-5 Stages in the surface development of gullies on a hillside
(Leopold et al., 1964)

2.3 Headward erosion

2.3.1 Conception

Headward erosion refers to the erosion occurring at a stream channel's source, which causes the source to migrate backward from the direction of stream flow, thus elongating the channel. It can also describe the widening of a canyon at its top edge due to erosion when sheets of water flow from a planar surface into the canyon, like those seen at Canyonlands National Park in Utah. When water sheets from an elevated surface enter a depression, they erode the top edge of the depression, either extending the stream channel backward (lengthening it) or widening the canyon along its top edge as water flows over it. Erosion occurring inside the canyon or beneath the canyon's edge—such as erosion from streamflow—is not considered headward erosion.

Headward erosion is a fluvial process that extends and enlarges a stream, valley, or gully at its source and increases its drainage basin. The processis shown in Fig. 2-6, the profile of the ditch head gradually changes from arc ab to arc AB. Streams erode rock and soil at their headwaters in the direction opposite to their flow. After a stream starts to erode backward, the process accelerates due to the steep gradient. As water erodes its path from the headwaters to its mouth, it tends to create a shallower path, intensifying erosion in the steepest parts. This process is headward erosion. Over time, continued headward erosion may enable a stream to break into an adjacent watershed, capturing drainage previously directed elsewhere.

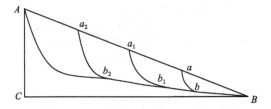

Fig. 2-6 Headward erosion

2.3.2 Stream types created by headward erosion

Headward erosion can form three types of streams: Insequent, subsequent, and obsequent/resequent streams. Insequent streams result from random headward erosion, typically from sheet flow over the land surface. Water accumulates in channels, increasing velocity and erosional power, which carves and extends gully heads. Subsequent streams arise from selective headward erosion, cutting away at less resistant rocks. Obsequent and resequent streams emerge over time in regions of insequent or subsequent streams. Obsequent streams are those that now flow in a direction opposite to their original pattern, while resequent streams are those that have similarly changed direction from their initial paths.

Fig. 2-7 illustrates the process of headward erosion in a river system. The upper part shows the initial stage, where erosion lengthens, deepens, and widens the river valley. Over time, as depicted in the lower part, these erosional processes continue to shape the landscape, forming a more pronounced and expansive canyon.

2.3.3 Drainage patterns created by headward erosion

Headward erosion shapes three major drainage patterns: dendritic, trellis, and rectangular/angular patterns. Dendritic patterns develop in homogenous terrains without structural control from underlying bedrock, characterized by randomly branching streams at acute angles. Trellis patterns occur where alternating weak and strong rocks exist, characterized by nearly parallel streams that form higher-angle branches. Rectangular and angular patterns involve tributaries branching at nearly right angles or exhibiting right-angle bends, typically forming in jointed igneous bedrock or sedimentary layers with intersecting faults.

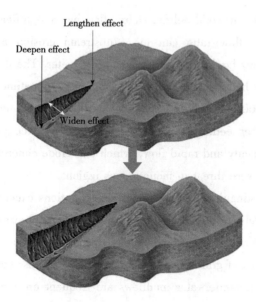

Fig. 2-7 Schematic diagram of river erosion

Additionally, four minor drainage patterns can arise: radial, annular, centripetal, and parallel patterns. Radial patterns flow outward from a central point, such as a volcanic cone or dome. Annular patterns form on domes of alternating weak and hard rock, resembling a bullseye from above due to differential erosion. Centripetal patterns direct water into a central location, like a sinkhole in karst limestone terrains. Parallel patterns are rare, forming in areas with unidirectional slopes or features, usually localized.

2.4 Debris flow

2.4.1 Conception

Debris flow is a type of torrent characterized by a high concentration of solid materials, such as sediment and stones. It erupts suddenly, has a short duration, and possesses significant destructive power. The volume of solid material often exceeds 25%, reaching up to 80% with a bulk density ranging from 1.3 to 2.3 t/m^3.

Debris flow exhibits a velocity gradient, showing a progressive change in velocity between the flow and the bottom of the channel. This feature differentiates it from collapses and landslides that involve fracture surfaces.

2.4.2 Features and behavior

Debris flow is a mixed erosional process involving both runoff erosion and gravity. It includes gravitational processes like collapses and landslides for material supply, along with the scouring effects of water flow, particularly the transport, deposition, impact, vibration, and abrasion of slurry rich in solid debris. Debris flow has a significant transport capacity. A 5% reduction in debris flow ratio can move upstream erosion materials to downstream locations, and it can carry boulders over 1 meter in diameter. The deposition of debris raises the downstream channel bed by about 2 meters annually, sometimes up to 5 meters, with impact forces between 200 to 800 N/m^2, peaking at 5000 N/m^2, and elevation rises of 3 to 5 times that of the "head" exceeding 10 meters. Debris flows produce ground tremors and noises due to their massive capacity and rapid flow, which can erode concrete or masonry by several centimeters, posing a severe threat in mountainous regions.

Debris flows are distinguished by sediment concentrations exceeding 40% to 50%, with the rest being water. "Debris" includes a wide range of sediment sizes, from microscopic clay particles to large boulders. Although media often refer to debris flows as mudflows, true mudflows primarily consist of grains smaller than sand and are less common on land compared to debris flows. However, underwater mudflows are frequent on continental margins, where they can generate turbidity currents. In forested areas, debris flows may also contain significant amounts of woody debris. Flows with solid concentrations between 10% and 40% are referred to as hyperconcentrated flows, differing from normal stream flows with lower sediment concentration.

Debris flows accelerate downhill via gravity, following steep mountain channels that emerge onto alluvial fans or floodplains. The flow's front, or "head" usually contains coarse materials like boulders and logs, creating high friction. Behind this is a lower-friction, mostly liquefied flow body rich in sand, silt, and clay, maintaining high pore-fluid pressures that enhance flow mobility. Some flows have a watery tail transitioning into hyperconcentrated flows. Debris flows typically move in pulses or surges, each with a distinct head, body, and tail.

Debris flow deposits are identifiable in the field and make up significant portions of alluvial fans and debris cones along mountain fronts. Exposed deposits often feature lobate forms with boulder-rich snouts, and lateral margins marked by boulder-rich levees. These natural levees form when more mobile, fine-grained debris pushes aside coarse materials, forming debris-flow heads. The presence of older levees helps estimate previous debris flow magnitudes, and the frequency of such flows can be determined by dating trees on the deposits—important data for land development in these areas. Ancient deposits revealed in outcrops are more challenging to recognize but are typically marked by poorly sorted sediments of varying shapes and sizes, distinguishing them from water-laid sediments.

Fig. 2-8 likely depicts a debris flow scenario. It shows a catchment area where intense rainfall or rapid snowmelt gathers, leading to a flow of water mixed with soil and rock down a steep channel. The channel gradient influences the speed and intensity of the debris flow. When the flow reaches flatter terrain, it spreads out, forming a depositional fan. The diagram highlights features such as the catchment internal relief, maximum runout distance, and lateral width of the fan, all relevant to understanding debris flow dynamics.

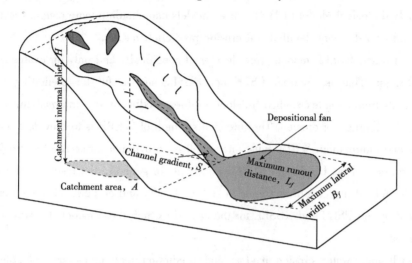

Fig. 2-8 Schematic diagram of typical debris flow basin

Extending reading: Water erosion models and prediction techniques

Water erosion models and prediction techniques are tools and methods used to assess and predict the process of water erosion. They are based on soil erosion equations and related physical processes, combined with factors such as climate, soil, vegetation, and topography, to provide quantitative estimates of water erosion risk. Here is a detailed explanation of water erosion models and prediction techniques:

The Revised Universal Soil Loss Equation (RUSLE) is one of the most commonly used erosion assessment models. It evaluates the following key factors: Rainfall energy (R), soil erodibility (K), slope length and steepness (LS), cover management (C), and support practices (P).

In addition to the RUSLE model, there are other soil erosion prediction techniques based on soil erosion equations, such as the Universal Soil Loss Equation (USLE) and the Modified Universal Soil Loss Equation (MUSLE). The USLE model is a precursor to the RUSLE model and is used for quantitatively assessing water erosion risk. The MUSLE model is an improvement over the USLE model, incorporating considerations for rainfall and slope variability, thereby enhancing the predictive accuracy of the model.

These prediction techniques typically require relevant Geographic Information System (GIS) data, including terrain data, soil data, weather data, vegetation data, etc., to simulate and predict water erosion processes.

In addition to equation-based models, more complex numerical simulation models are available for water erosion prediction. Some commonly used numerical simulation models include the Water Erosion Prediction Project (WEPP), River Erosion Model (REM), and Distributed Hydrological Model (DHM). These models can simulate more complex water erosion processes and provide more detailed soil erosion prediction results.

The USLE and RUSLE models were developed specifically and only for evaluating average annual soil loss: That is, neither USLE nor RUSLE are useful for predicting deposition, sediment yields from complex shaped hillslope profiles, sediment size information, or temporal and spatial distributions of erosion. The intended strength of WEPP is to provide a tool for such evaluations and information. WEPP must, nonetheless, provide accurate soil loss predictions for the simple cases such as USLE and RUSLE represent. In general, the results of these analyses show that the WEPP model follows trends in soil erodibility and cropping effects on erosion that are exhibited by the USLE and RUSLE for the cases in which those models follow known trends in existing data.

In conclusion, water erosion models and prediction techniques are valuable tools for assessing and predicting water erosion processes. They combine soil erosion equations and relevant physical processes, taking into account factors such as rainfall, soil, vegetation, and topography. They assist in the quantitative assessment of soil erosion risk and the development of appropriate soil conservation measures.

Source:
Boardman J, Favis-Mortlock D. Modelling Soil Erosion by Water, 1st ed. 1998.

Unit 3 Wind Erosion

3.1 Conception

Wind erosion is a type of erosion with wind as external force, that is the surface materials are displaced under the action of wind, which leads to the destruction and loss of the lithosphere or soil sphere. So wind erosion is a key link in the activity of wind-blown sand, which mainly occurs in arid and semi-arid regions. It is also a major source of land degradation, evaporation, desertification, harmful airborne dust, and crop damage—especially after being increased far above natural rates by human activities such as deforestation, urbanization, and agriculture.

Wind erosion is of two primary varieties: ①Deflation, where loose, fine-grained particles are picked up and carried away by the turbulent action of the wind; ②Abrasion, where surfaces are worn down as they are struck by airborne particles carried by wind.

Deflation also known as net wind erosion, when the wind blows through the surface, the loose sediment or bedrock weathering (sand material) on the surface is blown away due to the dynamic pressure of the wind, causing damage to the surface. During the process of deflation, the displacement of surface materials occurs due to the direct action of wind force, so deflation is also called fluid wind erosion. Deflation is very sensitive to the agglomeration of surface particles, so it generally occurs in dry and loose sandy surface, and the deflation is weaker in the surface with higher clay content and cementation. In the same event, the deflation intensity decreases with time. In the process of deflation, the deflation resistance of surface materials to will gradually increase.

Abrasion, also known as wind-sand flow erosion, refers to the surface damage and material displacement caused by the impact of moving sand particles on the surface when wind-sand flow (sand carrying airflow) blows through the surface. The intensity of abrasion is generally much greater than that of deflation. The wind tunnel experiment shows that the abrasion intensity is

4~5 times that of deflation intensity when subjected to sand carrying wind erosion at the same wind speed. Because the density of sand is more than 2000 times that of density of air, when it moves at the same speed as the air flow, the energy is very large. The impact of saltating sand particles on the loose surface will cause more particles to enter the airflow or rebound back into the airflow, continuously accelerating in the airflow to obtain more energy and then impact the surface. In this repeated process, more wind motion is transmitted to the surface, increasing the intensity of wind erosion. When saltating sand particles collide with a relatively solid surface, the first thing is that the impact of sand particles damages the surface, producing loose sand particles. During the abrasion process, the impact particles are the transmitters of wind energy and the direct carriers of wind erosion energy, so they are also called impact wind erosion. Once a aeolian sand movement occurs, abrasion is the main form of wind erosion and becomes the main driving force for shaping wind eroded landforms.

3.2 Transport

In general, particles are transported by winds in three ways (Fig. 3-1): ①Suspension, where very small and light particles are lifted into the air by the wind, and are often carried for long distances; ②Saltation (skipping or bouncing), where particles are lifted a short height into the air, and bounce and saltate across the surface of the soil; ③Creeping (rolling or sliding) where larger, heavier particles slide or roll along the ground.

Particles may be held in the atmosphere in suspension. Upward currents of air support the weight of suspended particles and hold them indefinitely in the surrounding air. Suspension generally affects small particles ("small" means 70 μm or less for particles in air). For these particles, vertical drag forces due to turbulent fluctuations in the fluid are similar in magnitude to the weight of the particles. These smaller particles are carried by the fluid in suspension, and advected downstream. The smaller the particle, the less important the downward pull of gravity, and the longer the particle is likely to stay in suspension.

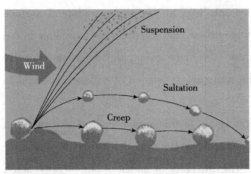

Fig. 3-1 Three ways of particles transport

In geology, saltation (from Latin saltus, "leap") is downwind movement of particles in a series of jumps or skips. Once the wind speed reaches a certain critical value, termed the impact or fluid threshold, the drag and lift forces exerted by the fluid are sufficient to lift some particles from the surface. These particles are accelerated by the fluid, and pulled downward by gravity, causing them to travel in roughly ballistic trajectories. Saltation normally lifts sand-size particles no more than one centimeter above the ground and proceeds at one-half to one-third the speed of the wind. It occurs when loose materials are removed from a bed and carried by the fluid, before being transported back to the surface. If a particle has obtained sufficient speed from the acceleration by the fluid, it can eject, or splash, other particles in saltation, which propagates the process. Depending on the surface, the particle could also disintegrate on impact, or eject much finer sediment from the surface. In air, this process of saltation bombardment creates most of the dust in dust storm.

Recent study finds that saltating sand particles induces a static electric field by friction. Saltating sand acquires a negative charge relative to the ground which in turn loosens more sand particles which then begin saltating. This process has been found to double the number of particles predicted by theory. This is significant in meteorology because it is primarily the saltation of sand particles which dislodges smaller dust particles into the atmosphere.

At low fluid velocities, loose material rolls downstream, staying in contact with the surface. This is called creep or reptation. Here the forces exerted by the fluid on the particle are only enough to roll the particle around the point of contact with the surface. Alternatively, a saltating grain may hit larger grains that are too heavy to hop, but that slowly creep forward as they are pushed by saltating grains.

Saltation is responsible for the majority (50%~70%) of wind erosion, followed by suspension (30%~40%), and then surface creep (5%~25%). Silty soils tend to be the most affected by wind erosion; silt particles are relatively easily detached and carried away.

3.3 Deposition

Wind-deposited materials hold clues to past as well as to present wind directions and intensities. These features help us understand the present climate and the forces that molded it. Wind-deposited sand bodies occur as sand sheets, ripples, and dunes.

Sand sheets are flat, gently undulating sandy plots of sand surfaced by grains that may be too large for saltation. They form approximately 40% of aeolian depositional surfaces. The Selima Sand Sheet in the Eastern Sahara Desert, which occupies 6×10^4 km^2 in Southern Egypt and Northern Sudan, is one of the earth's largest sand sheets. The Selima is absolutely flat in a few places; in others, active dunes move over its surface.

Wind blowing on a sand surface ripples the surface into crests and troughs whose long axes are perpendicular to the wind direction. The average length of jumps during saltation corresponds to the wavelength, or distance between adjacent crests, of the ripples. In ripples, the coarsest materials collect at the crests causing inverse grading. This distinguishes small ripples from dunes, where the coarsest materials are generally in the troughs. This is also a distinguishing feature between water laid ripples and aeolian ripples.

Accumulations of sediment blown by the wind into a mound or ridge, dunes have gentle upwind slopes on the windward side. The downwind portion of the dune, the lee slope, is commonly a steep avalanche slope referred to as a slipface. Dunes may have more than one slipface. The minimum height of a slipface is about 30 cm. Wind-blown sand moves up the gentle upwind side of the dune by saltation or creep. Sand accumulates at the brink, the top of the slipface. When the buildup of sand at the brink exceeds the angle of repose, a small avalanche of grains slides down the slipface. Grain by grain, the dune moves downwind.

In addition, most of the dust carried by dust storms is in the form of silt-size particles. Deposits of this windblown silt are known as loess. The thickest known deposit of loess, 335 m, is on the Loess Plateau in China. This very same Asian dust is blown for thousands of miles, forming deep beds in places as far away as Hawaii. In Europe and in the Americas, accumulations of loess are generally 20~30 m thick. The soils developed on loess are generally highly productive for agriculture.

3.4　Aeolian landform

Aeolian landforms are features of the earth's surface produced by either the erosive or constructive action of the wind. This process is not unique to the earth, and it has been observed and studied on other planets, including Mars. According to the formation mechanism, aeolian landforms can be divided into two major types wind erosion landforms and wind deposition landforms and they can be further divided into several subtypes.

3.4.1　The types of wind erosion landforms

Wind erosion landform types mainly include yardangs, mushroom rocks and blowouts etc.

(1) Yardang

"Yardang" is a Turkish word first introduced by Sven Hedin in the early 20th century. A yardang is a streamlined protuberance carved from bedrock or any consolidated or semi-consolidated material by the dual action of wind abrasion by dust and sand, and deflation which is the removal of loose material by wind turbulence. Yardangs become elongated features typically three or more times longer than wide, and when viewed from above, resemble the hull of a boat. Facing the wind is a steep, blunt face that gradually gets lower and narrower toward the lee end. Yardangs are formed by wind erosion, typically of an originally flat surface formed from areas of harder and softer material. The soft material is eroded and removed by the wind, and the harder material remains. The resulting pattern of yardangs is therefore a combination of the original rock distribution, and the fluid mechanics of the air flow and resulting pattern of erosion.

Yardangs can be found in most deserts across the globe. Depending upon the winds and the composition of the weakly indurated deposits of silt and sand from which they are carved, yardangs may form very unusual shapes, including deflation mushroom, deflation column and deflation residual hill etc. some even resembling humans or animals(Fig. 3-2).

Fig. 3-2 Yardang in the shape of a peacock

The sizes of Yardangs vary from a few centimeters to several kilometers, and they are divided into three different categories: mega-yardangs, meso-yardangs, and micro-yardangs. Mega-yardangs can be several kilometers long and hundreds of meters high and are found in arid regions with strong winds; meso-yardangs are generally a few meters high and 10~15 m long and are commonly found carved in semiconsolidated playa sediments and other soft granular materials; and micro-yardangs are only a few centimeters high.

A large concentration of mega-yardangs are found near the Tibesti Mountains in the central Sahara. There is a famous yardang at hole in the rock in Papago Park in Phoenix, Arizona, a rock formation with a roughly circular hole in it. Another yardang in Arizona is Window Rock, near the town of Window Rock. It is a 60 m sandstone hill with a very large circular hole in the middle of it. Some geologists have suggested that the Great Sphinx of Egypt is an augmented yardang. Pictures from Mars show that the yardang ridges occur on a massive scale there; some individual ridges are tens of kilometers long with intervening valleys nearly 1 km wide.

(2) Mushroom rock

A mushroom rock, rock pedestal, or gour is a typical mushroom-shaped landform that is formed by the action of wind erosion(Fig. 3-3). These rocks usually found in desert areas, and form over thousands of years when wind erosion of an isolated rocky outcrop progresses at a different rate at its bottom than at its top. Abrasion by wind-borne grains of sand is most prevalent within the first 0.9 m(3 feet) above the ground, causing the bases of outcrops to erode more rapidly than their tops. Occasionally, the chemical composition of the rocks can be an important factor; if the upper part of the rock is more resistant to chemical erosion and weathering, it erodes more slowly than the base. For example, erosion attributed to chemical weathering at the base of the rock due to the collection of dew near the surface. The layer of softer rock is more readily eroded, leading to the formation of a depression or blowout.

Fig. 3-3　A mushroom rock

(3) Blowout

Blowouts also called deflation basins, are sandy depressions in a sand dune ecosystem caused by the removal of sediments by wind(Fig. 3-4), they are generally small, but may be up to several kilometers in diameter. Commonly found in coastal settings and arid margins, blowouts tend to form when wind erodes into patches of bare sand on stabilized vegetative dunes. Generally, blowouts do not form on actively flowing dunes due to the fact that they need to be bound by some extent, such as plant roots. These depressions usually start on the higher parts of the stabilized dunes on the account that desiccation and disturbances are more considerable which allows for greater surface drag and sediment entrainment when sand is bare. Most of the time, exposures become quickly re-vegetative before they could become blowouts and expand; however, when the opportunities are given, wind erosion can lower the exposure surface and create a tunneling affect, which increases the wind speed. The depression may continue until it hits a non-erodible substrate or morphology limits it. The eroded substances climb the steep slopes of the depression and become deposited on the downwind side of the blowout which can form a dune that covers vegetation and lead to a

Fig. 3-4 A blowout

larger area of depression; a process that helps create parabolic dunes.

The scientific community mostly utilizes two types of blowouts: trough and saucer. Although there is no obvious reason why one type is formed rather another in a particular region, saucer blowouts generally have semicircular and saucer shapes while trough blowouts have more elongated shapes with deep deflation basins and steeper slopes. Nevertheless, both types of blowouts have structures that can affect wind flow within the basin.

In troughs, the structure's topography can accelerate flows and form jets that result in maximum erosion along the deflation basin floor and laterally expand the slopes of the blowout. Additionally, when the wind flows on top of the blowout's lateral walls, sediment transport is at its maximum in the middle axis of the trough depositional lobe, leading to formation of a parabolic dune.

Saucer blowouts indicate a deceleration of wind flow along the deflation basin as the structure widens over time by reversing flows eroding the sides and expanding upwind. Due to rapid deceleration, saucers tend to form short, wide, radial depositional slopes. When wind flow enters a saucer shape blowout, the wind speed decreases upon entering the blowout and accelerates at the downwind side of the formation. A zone of separation develops along the lee slope as the wind enters the blowout and decrease in speed, yet it accelerates again as it re-attaches at the basin and flow up to the depositional lobe, where sand becomes evacuated.

Even though they are more influences blowout structures have on their morphology, both types basically tend to have deflation basins eroded until they reach their non-erodible base level. A study conducted by Hesp (1982) indicates that depositional length is not correlated with the eroded depth but rather the blowout width. In other words, as the depositional lobe increases, the blowout width also increases by a ratio of 1∶2~1∶3 in saucer blowouts and 1∶4 in trough blowouts.

(4) Ventifacts

Ventifacts are formed by sand blast and dust abrasion of pebbles and they range in diameter from a few centimeters to several meters. Ventifacts are the most common small-scale wind erosion landform in arid deserts, especially in the vast gravel desert (conglomerate desert). Of course, They are not limited to modern arid desert areas, but can also occur in ancient and modern periglacial regions. The gravel in the vast conglomerate desert, after long-term erosion by wind-sand flow, can become clear angular, smooth surface of the ventifacts (Fig. 3-5).

Fig. 3-5 A ventifact

Depending on the number of edges, there are three types of ventifacts: Single-edged, three-edged and multi-edged ventifact, but three-edged ventifact is the most common. The reason for this is that some of the gravel that protrudes from the surface, after long-term grinding by directional wind-sand flow, forms a polished surface (wind erosion surface); in the future, due to changes in wind direction or the overturning and reorientation of gravel, another polished surface is formed; there are sharp edges between the faces, forming a ventifact. The number of edges depends not only on changes in wind direction and the number of times the gravel flips, but also on the shape of the original gravel, as the ventifact is formed by polishing on the basis of the original shape of the gravel. Therefore, the formation of ventifacts generally requires the following conditions: Strong winds and open ground favorable for wind action; suitable sand particles are provided for sufficient wind erosion.

3.4.2 The types of wind deposition landforms

Wind deposition landform mainly refers to sand dunes, they are formed by the interaction of sand materials and wind. Sand dunes can have a negative impact on humans when they encroach on human habitats. Sand dunes move via a few different means, all of them helped along by wind (Fig. 3-6). One way that dunes can move is by saltation, where sand particles skip along the ground like a bouncing ball. When these skipping particles land, they may knock into other particles and cause them to move as well, in a process known as creep. With slightly stronger winds, particles collide in mid-air, causing sheet flows. In a major dust storm, dunes may move tens of meters through such sheet flows. Also as in the case of snow, sand avalanches, falling down the slipface of the dunes—that face away from the winds—also move the dunes forward.

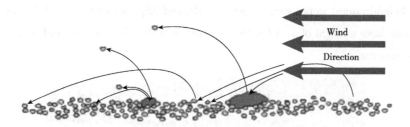

Fig. 3-6 Formation process of dunes
(Sand hitting sand is more likely to stick; sand hitting a more coherent
surface is more likely to bounce. This exacerbating feedback
loop helps sand accumulate into dunes)

The sand mass of dunes can move either windward or leeward, depending on if the wind is making contact with the dune from below or above its apogee. If wind hits from above, the sand particles move leeward. If sand hits from below, sand particles move windward. The leeward flux of sand is greater than the windward flux. Further, when the wind carrying sand particles when it hits the dune, the dune's sand particles will saltate more than if the wind had hit the dune without carrying sand particles.

According to morphological and genetic classification, sand dunes are divided into five major types: Barchan, linear, dome, star and parabolic dunes.

All these dune shapes may occur in three forms: Simple (isolated dunes of basic type), compound (lager dunes on which smaller dunes of same type form), and complex (combinations of different types). Simple dunes are basic forms with the minimum number of slipfaces that define the geometric type. Compound dunes are large dunes on which smaller dunes of similar type and slipface orientation are superimposed. Complex dunes are combinations of two or more dune types. A crescentic dune with a star dune superimposed on its crest is the most common complex dune. Simple dunes represent a wind regime that has not changed in intensity or direction since the formation of the dune, while compound and complex dunes suggest that the intensity and direction of the wind has changed. According to the statistical data by Fryberger et al. (1981), about 46.6% of modern sand sea is covered by compound or complex dunes.

(1) Barchan dunes

①Simple barchan dunes: Simple barchan dunes are crescent-shaped mounds which width generally greater than their Length (Fig. 3-7). The lee-side slipfaces are on the concave sides of the dunes. These dunes mainly occur in deserts with uni-directional wind and meager sand supply and they occupy a very small area in the global sand sea. Their two wings extend downwind with a slip face. Their windward slopes range from 2°~15° and slip faces are 30°~35°. Most of simple barchan dunes have a height of 1~10 m, the ratio between their height and width is 1 : 10, and two wings are symmetric.

Fig. 3-7 A simple barchan dune

Some types of barchan dunes move more quickly over desert surfaces than any other type of dune. A group of dunes moved more than 100 m/a between 1954 and 1959 in Ningxia, and similar speeds have been recorded in the Western Desert of Egypt. The largest barchan dunes on the earth, with mean crest-to-crest widths of more than 3 km, are in Taklamakan Desert.

②Barchanoid ridges: Abundant barchan dunes may merge into barchanoid ridges, which then grade into linear (or slightly sinuous) transverse dunes (Fig. 3-8), so called because they lie transverse, or across, the wind direction, with the wind blowing perpendicular to the ridge crest. Barchanoid ridges occupy about 40% of the total area of the world sand sea, they are widely distributed in various sand sea and. Their heights range from 3~10 m and spacings vary between 100~400 m, the slope from the base of windward slope to the mid-upslope varies from 2°~3° to 10°~12°. Their crest lines are bended. Some scholars called their projective part tongue or crescentic sand ridge. Complex barchanoid ridges are 20~80 m in height and a few hundred meters to tens of kilometers in length, with simple barchan dunes superimposed on their windward slope.

Fig. 3-8 The barchanoid ridges

(2) Linear dunes

Linear dunes occupy about 50% of the global sand sea area, but their areas are different in different regions, for example, in the Kalahari Sand Sea it occupies 85%~90% of its total area and in the Gran Desieto it accounts for 1%~2%.

①Simple linear dunes: Simple linear dunes have two forms: the first one has flat and straight crest line and the surface is covered by vegetation, as can be seen in the Southwest Kalahari Desert; the second one has bend crest line and sharp crest, they mainly occur in the Arabian peninsula and are also known as "seif". Simple linear dunes are generally 2~35 m in height, their spacing is 200~450 m, width 150~250 m, length 20~25 km and seldom exceeds 200 km. Some scholars found that there is a close correlation between the height and spacing of simple linear dunes.

The formation mechanism of the two forms of linear dunes differs greatly. First, the first form of linear dune resulted from deflation and cutting through of the nose of parabolic dunes and vegetation play an important role in the formation of such dunes.

The formation mechanism of the second type of linear dune, the seif dunes, is debated. Bagnold, in The Physics of Blown Sand and Desert Dunes, suggested that some seif dunes form when a barchan dune moves into a bidirectional wind regime, and one arm or wing of the crescent elongates. Others suggest that seif dunes are formed by vortices in a unidirectional wind. In the sheltered troughs between highly developed seif dunes, barchans may be formed, because the wind is constrained to be unidirectional by the dunes.

Seif dunes are common in the Sahara. They range up to 300 m in height and 300 km in length. In the southern third of the Arabian Peninsula, a vasterg, called the Rub'al Khali or Empty Quarter, contains seif dunes that stretch for almost 200 km and reach heights of over 300 m (Fig. 3-9).

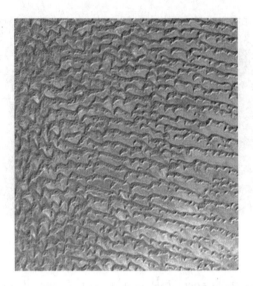

Fig. 3-9 Rub'al Khali sand dunes imaged by Terra
(Most of these dunes are seif dunes: Their origin from barchans is
suggested by the stubby remnant "hooks" seen on many of the dunes;
Wind would be from left to right)

②Compound linear dunes: Compound linear dunes consist of 2~4 ridges with bend crest line, they have one main crest line, their tops show a star-like shape, with barchan dunes occurred on both sides. They are common in the Namib Sand Sea. Generally, there are 3~5 main dunes running parallel with each other. Furthermore, their bend and sharp sand crests are 5~10 m high. Such dunes are 50~800 m wide, 25~50 m high and 1200~2000 apart each other. Their crests consist of coarse sand, there are "Y" junctions and the interdune flats are often covered by sand.

③Complex linear dunes: Complex linear dunes are fairly common. Their height may reach 50~170 m, spacing 1600~2800 m and mean spacing 2000 m or so. There is a good correlation between their spacing and height. Their crest lines generally are bend, leeward slope is 32° and there are 2~10 m high and 50~200 m wide secondary dunes superimposed on the main dunes.

(3) Dome dunes

①Simple dome dunes: Dome dunes are also called round dunes. Their morphological characteristics are as follows: the slopes on both sides are relatively symmetrical, there is no obvious curved sand fall slope, the length and width are roughly equal, and the plan is round or oval, like steamed bun. Dome dunes are rare and occur at the far upwind margins of sand seas. They are generally low. A typical dome dune in the White Sands region of New Mexico, USA, with a length of 137 m, a width of 128 m, and a height of 5.5 m.

②Compound dome dunes: Some dome dunes have secondary dunes stacked layer by layer, becoming compound dome dunes, which are relatively tall. Generally, they present scattered and irregularly distribution individually, some areas are also connected, but still maintain the morphological characteristics of each dome. These compound dome dunes are distributed in the north of Taklamakan Desert, south of the old riverbed of Tarim River, and southwest of Ulanbuh Desert, with a height of 40~60 m. Dome dunes are also distributed in the northeast and northwest of the western desert in Algeria, the An Nafud and other places in the Arabian Peninsula, and most of them belong to tall composite dome sand dunes. Dome sand dunes are also distributed in the northeast and northwest of Algeria's Western Desert, the An Nafud and other places in the Arabian Peninsula, and most of them belong to tall compound dome dunes.

(4) Star dunes

Radially symmetrical, star dunes are pyramidal sand mounds with slipfaces on three or more arms that radiate from the high center of the mound (Fig. 3-10). Their arms are not necessarily balanced. They tend to accumulate in areas with multidirectional wind regimes. In many areas one of their arms has seasonal changes due to the changes in wind direction. There may be secondary barchan dunes or reversing dunes in the interdune flats. Star dunes grow upward rather than laterally.

Fig. 3-10 An isolated star dune

About 8.5% of the global sand sea is covered by star dunes. They dominate the Grand Erg Oriental of the Sahara. In other deserts, they occur around the margins of the sand seas, particularly near topographic barriers. Their spacing ranges from 150~5000 m, generally 1000~2400 m. Star dunes are the largest dune type in the world, their mean height is 117 m and varying in different regions. In the Southeast Badain Jaran Desert of China, the star dunes are up to 500 m tall and may be the tallest dunes on the earth.

(5) Parabolic dunes

U-shaped mounds of sand with convex noses trailed by elongated arms are parabolic dunes (Fig. 3-11). These dunes often occur in semiarid areas where the precipitation is retained in the lower parts of the dune and underlying soils. The stability of the dunes was once attributed to the vegetative cover but recent research has pointed to water as the main source of parabolic dune stability. These type of sand dunes typically occur where vegetation is locally destroyed and is formed by the downwind extension of blowouts. Their arms are fixed by vegetation and point toward upwind direction. The paired arms of most parabolic dunes may reach 1~2 km long and 10~70 m high, but the nose of the dune is lower in height, where vegetation stops or slows the advance of accumulating sand. With the retreat of their U-shaped nose and the extension of their paired wings, sand supply to the nose become less and less and finally the nose is cut through to form linear dunes.

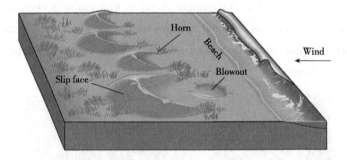

Fig. 3-11 Schematic of coastal parabolic dunes

Parabolic dunes, like crescent dunes, occur in areas where very strong winds are mostly unidirectional. Although these dunes are found in areas now characterized by variable wind speeds, the effective winds associated with the growth and migration of both the parabolic and crescent dunes probably are the most consistent in wind direction. Unlike crescent shaped dunes, their crests point upwind. The bulk of the sand in the dune migrates forward.

Due to the fact that the arms of parabolic dunes are covered with grass, shrubs, and trees, coppice dunes gradually form over time. They have a height of less than 1 m but in some areas they may reach a height of 3.5 m or so. Owing to low sand flux they generally have along development history.

> **Extending reading: Measures to prevent and control aeolian sand drift**
>
> Severe environmental problems caused by windblown sand events are one of the important issues affecting the safe operation of transportation and economic development in the desert regions; these issues will become increasingly prominent with the strategic promotion of

development in Western China. To mitigate problems associated with aeolian sand drift, multiple sand control measures can be applied; these include the blocking, fixing, transport and guiding of sand. Among these, sand blocking and sand fixing measures are the most effective and most commonly used approaches to prevent and control the sand hazard, and can include sand fences, sand-proof dikes, wind-breaks, shelterbelts, checkerboard fences, chemical reagents and others.

A sand fence is regarded as a typical sand blocking measure in the windblown sand project, and functions to intercept and prevent the movement of windblown sand by reducing the wind velocity as much as possible. Windblown sand particles are deposited in the vicinity of sand fence, thereby cutting off the sand source and greatly reducing the effect of windblown sand; this can effectively prevent sand damage to railways and highways in desert areas. The blocking effect of the sand fence depends on its specifications, such as porosity, height, location and thickness.

Many experimental studies have demonstrated that the porosity of the fence is generally the most significant factor controlling the aerodynamic performance and blocking effect of a fence with a given height.

Many experts measured mean wind velocity, root mean square velocity (RMS velocity), velocity fluctuations, energy spectra and shelter efficiency of model fences with different porosities, concluding that the fence with 30%~50% porosity was most effective in leeward wind reduction; In summary, the optimal porosity range of sand fences is considered to be 30%~50%.

The height of the fence is generally considered to be another important factor affecting the blocking effect of the fence: if the sand fence is too low, it will be buried quickly, necessitating frequent construction which is neither economical nor convenient; if the fence is too high, wind pressure is greater and the construction will be more difficult. The sheltered area is proportional to the height of the fence, and the protection range is usually expressed as a multiple of the fence height.

Planting vegetation also helps to stabilize desertification environment. Here are some related studies:

A 1998 study published in *Earth Surfaces Processes and Landforms* investigated the relationship between vegetative cover on sand surfaces with the rate of sand transport. It was found that sand flux decreased exponentially with vegetation cover. This was done by measuring plots of land with varying degrees of vegetation against rates of sand transport. The authors contend that this relationship can be utilized to manipulate rates of sediment flux by introducing vegetation in an area or to quantify human impact by recognizing vegetation loss's effect on sandy landscapes.

A three-year quantitative study on the effects of vegetation removal on wind erosion found that the removal of grasses in an aeolian environment increased the rate of soil deposition. In the same study, a relationship was shown between decreasing plant density with decreasing soil nutrients. Similarly, horizontal soil flux across the test site was shown to increase with increasing vegetation removal.

A 2011 study published in *Catena* examined the effect of vegetation on aeolian dust accumulation in the semiarid steppe of Northern China. Using a series of trays with different vegetation coverage and a control model with none, the authors found that an increase in vegetation coverage improves the efficiency of dust accumulation and adds more nutrients to the environment, particularly organic carbon. Two critical point were revealed by their data: ①The efficiency of trapping dust increases slowly above 15% coverage, and decreases rapidly below 15% coverage; ②At around 55%~75% coverage, dust accumulation reaches a maximum capacity.

Source:

Chen G, Wang W, Sun C, et al. 3D numerical simulation of wind flow behind a new porous fence[J]. Powder Technology, 2012, 230: 118-126.

Vigiak O, Sterk G, Warren A, et al. Spatial modeling of wind speed around windbreaks[J]. Catena, 2003, 52 (3-4): 273-288.

Unit 4　Mass Erosion

4.1　Overview

Under the direct influence of gravity, mass movement occurs and manifests as downslope movement of materials such as sediments, rocks, plants et al. The material is transported from higher elevations to lower elevations, where streams or glaciers can pick it up and move to even lower elevations. It is not a type of erosion in itself, but when flowing water is present, the loosened material may be transported from the site, which is a type of soil erosion. The site from which the mass of soil moved often has a high potential for erosion, especially by water.

Some mass movement processes act very slowly, others occur suddenly, often with disastrous consequences. Any perceptible down-slope movement of rock or sediment is referred to in general terms as a landslide. It can be classified such that it reflects the mechanism responsible for the movement and the velocity at which the movement occurs.

Slumping occurs at steep hillsides, along distinct fracture planes, in some cases favored by water, often within materials such as clay. Once released, slumps may move rapidly downhill. They will often show a spoon-shape topographic depression where the material has begun to slide downhill. Surface creep is the slow downhill movement of soil and rock debris. It is usually not perceptible except through extended observation. The term can also describe the rolling of soil particles by wind along soil surface.

4.2　Sinkholes

Sinkholes occur naturally, especially where there is abundant rainfall, and the rock beneath the surface soil is limestone. In a landscape where limestone sits underneath the soil, water from rainfall collects in cracks in the stone. These cracks are called joints. Slowly, as the limestone dissolves and is carried away, the joints widen until the ground above them becomes unstable and collapses. The collapse often happens very suddenly and without very much warning. Water collects in these collapsed sections, forming sinkholes.

Sinkholes also form when the roofs of caves collapse. Sinkholes are often funnel-shaped, with the wide end open at the surface and the narrow end at the bottom of the pool. Sinkholes vary from shallow holes about 1 meter deep, to pits more than 50 m deep. Water can drain through a sinkhole into an underground channel or a cave. When mud or debris plugs one of these underground caves, it fills with water to become a lake or a pond.

The land surrounding the Dead Sea in the Middle East is prone to sinkholes because of the prevalence of rock salt, which is easily dissolved by water. Tourists who are unaware of sinkholes and even scientists studying sinkholes have been injured by falling into them.

People can create sinkholes when building roads, aquifers, or other types of construction. Altering land in these ways can weaken the underlying rock and make it more susceptible to sinkholes. Sinkholes can open up in the middle of busy streets or in neighborhoods, especially during heavy rainfall.

Sinkholes in the plain areas of Southern Italy developed in different geological settings, which may be schematically classified as: alluvial plains; intramontane Apennine basins; fluvial valleys; coastal plains. Sinkholes mainly develop where the following factors are present: alluvial and/or volcaniclastic sediments, dominated by sand and silt facies and with a shallow water table. Alternating layers with different permeability, and the presence of multi-layered aquifers, seem to be the main predisposing factors in more than one type of setting (i.e. alluvial plains and fluvial valleys). In addition to karst processes which may affect the buried bedrock where suitable rocks are present at depth, groundwater circulation and water table fluctuations may cause internal erosion and the formation of cavities in the overburden, whose upward propagation eventually leads to sinkhole development.

4.3 Earthflows

Earthflows are the downslope movement (a slip or a slide) of soil on a relatively steep hillside. The movements occur after prolonged rainy periods when the soil becomes saturated, often where downward percolation of water is impeded by subsurface conditions such as impervious rock layers or dense clays. Earth flows result in loosening of the surface soil, which may contribute to erosion by water during the ongoing or a subsequent rainstorm.

Earthflows are one of the most fluid types of mass movements. Earthflows occur on heavily saturated slopes like mudflows or a debris flow. Though earthflows are a lot like mudflows, overall they are slower and are covered with solid material carried along by flow from within. Earthflows are often made up of fine-grained materials so slopes consisting of clay and silt materials are more likely to create an earthflow.

As earthflows are usually water-dependent, the risk of one occurring is much higher in humid areas especially after a period of heavy rainfall or snowmelt. The high level of precipitation, which saturates the ground and adds water to the slope content, increases the pore-water pressure and reduces the shearing strength of the material. As the slope becomes wet, the earthflow may start as a creep downslope due to the clay or silt having less friction. As the material is increasingly more saturated, the slope will fail, which depends on slope stability. In earthflows, the slope does not fail along a clear shear plane and is instead more fluid as the material begins to move under the force of gravity as friction and slope stability is reduced.

Earthflows can have sudden impacts on the amount of sediment that is deposited into a river system, which can have effects on the life in and around the river itself. They can also cause damage to roads and constructions built near the slope. One of the best mitigation techniques to avoid serious earthflow and landslide damage is properly draining the slope of water, especially in places of high levels of precipitation.

4.4 Landslides

Landslides, also known as landslips, are several forms of mass wasting that may include a wide range of ground movements, such as rockfalls, shallow or deep-seated slope failures, mudflows, and debris flows. Landslides occur in a variety of environments, characterized by either steep or gentle slope gradients, from mountain ranges to coastal cliffs or even underwater, in which case they are called submarine landslides.

Gravity is the primary driving force for a landslide to occur, but there are other factors affecting slope stability that produce specific conditions that make a slope prone to failure. In many cases, the landslide is triggered by a specific event (such as a heavy rainfall, an earthquake, a slope cut to build a road, and many others), although this is not always identifiable. Landslides are frequently made worse by human development (such as urban sprawl) and resource exploitation (such as mining and deforestation). Land degradation frequently leads to less stabilization of soil by vegetation. Additionally, global Warming caused by climate change and other human impact on the environment, can increase the frequency of natural events (such as extreme weather) which trigger landslides. Landslide mitigation describes the policy and practices for reducing the risk of human impacts of landslides, reducing the risk of natural disaster.

Landslides occur when the slope (or a portion of it) undergoes some processes that change its condition from stable to unstable. This is essentially due to a decrease in the shear strength of the slope material, an increase in the shear stress borne by the material, or a combination of the two. A change in the stability of a slope can be caused by a number of factors, acting together or alone.

Landslide hazard analysis and mapping can provide useful information for catastrophic loss reduction, and assist in the development of guidelines for sustainable land-use planning. The analysis is used to identify the factors that are related to landslides, estimate the relative contribution of factors causing slope failures, establish a relation between the factors and landslides, and to predict the landslide hazard in the future based on such a relationship. The factors that have been used for landslide hazard analysis can usually be grouped into geomorphology, geology, land use/land cover, and hydrogeology. Since many factors are considered for landslide hazard mapping, GIS is an appropriate tool because it has functions of collection, storage, manipulation, display, and analysis of large amounts of spatially referenced data which can be handled fast and effectively. Cardenas reported evidence on the exhaustive use of GIS in conjunction of uncertainty modelling tools for landslide mapping. Remote sensing techniques are also highly employed for landslide hazard assessment and analysis. Before and after aerial photographs and satellite imagery are used to gather landslide characteristics, like distribution and classification, and factors like slope, lithology, and land use/land cover to be used to help predict future events. Before and after imagery also helps to reveal how the landscape changed after an event, what may have triggered the landslide, and shows the process of regeneration and recovery.

Using satellite imagery in combination with GIS and on-the-ground studies, it is possible to generate maps of likely occurrences of future landslides. Such maps should show the locations of previous events as well as clearly indicate the probable locations of future events. In general, to predict landslides, one must assume that their occurrence is determined by certain geologic factors, and that future landslides will occur under the same conditions as past events. Therefore, it is necessary to establish a relationship between the geomorphologic conditions in which the past events took place and the expected future conditions.

Extending reading: Human impacts on the karst environment

Earth's landscapes are the result of a series of processes that may act continuously or during discrete events occurring with different temporal frequencies. Landscapes have been shaped, destroyed, and rebuilt again over geological times, in a natural cycle. In recent times, besides the complex suite of natural processes, the human factor has started playing an important role interfering directly (e. g. urban development) and indirectly (e. g. global warming) with the environment. The assessment of natural and anthropogenic impacts on the environment is of crucial importance for reducing the negative effects and promoting sustainable development. These assessments are common practice and are generally based on a set of biological and physical-chemical indicators.

In the past, notwithstanding the rugged topography and the paucity of surface water, humans tended to live close to karst areas, where the majority of springs are located. The increase in human population has progressively led people to occupy more karst areas and build new settlements and infrastructures.

Karst environments and the groundwaters associated with the mare highly vulnerable and sensitive to human alterations. The FAO (Food and Agriculture Organization of the United Nations) forecasts that before 2025 at least 80% of the demand of drinkable water in the Mediterranean Basin will be derived from karst aquifers. This highlights the urgent need of adequately protecting this resource from actions that may have adverse effects on its quality and availability. Protection of karst environments is a mandatory step to maintain, safeguard and transmit its extreme richness and biodiversity to future generations. This is a challenging task due to the increasing impact posed by human activities on the karst environments.

Especially in lowland karst, due to the scarce relief and the subdued features, many landforms may be lost due to anthropogenic activities such as intense quarrying, stone clearing and crushing practices. Land use changes also results in degradation of the epi karst, which, even in areas where it has a reduced thickness, provides a vital function for karst ecosystems, controlling runoff infiltration.

Further, the expansion of urban areas (including communication routes and industrial facilities) in karst is leading to an increasing number of pollution events, with severe consequences on the karst ecosystems and the quality of groundwater. The situation is even more exacerbated in post-conflict scenarios, as experienced for instance in karst areas in the Balkans. Actions must therefore be undertaken to assess the negative impacts of the increasing pressure on the fragile karst environment and minimize them.

Tourist exploitation in karst may have high impact if not reached through a careful evaluation of the deriving effects. Opening of show caves, for instance, must consider and respect the "visitor capacity", that is "that flow of visitors into a defined cave that confines the changes in itsmain environmental parameterswithin the natural ranges of their fluctuation".

Environmental Impact Assessments (EIA) of anthropogenic activities have become a mandatory requisite during the pre-approval stage of projects not only in Europe and America, but also in many other developed countries. These EIAs are often based on general indices that take into account the environmental, social and economic impacts, without taking into consideration the peculiarities of specific landscapes like those of karst terrains.

The protection of the karst environments, due to their intrinsic vulnerability, unique hydrological behavior, and exceptional subterranean ecosystems, requires specific approaches and measures. In karst, it is preferable to undertake vulnerability assessments startingwith a hydrogeological EIA, given the direct relationships between the surface and subsurface environments and the peculiar hydrologic regime. Subsequently, other fields of research should be investigated, depending on the study area.

The intrinsic vulnerability of karst aquifers to pollution may be assessed through the hydrogeological and geological parameters that determine the sensitivity of groundwater to contamination by human activities, and is independent of the nature of the contaminants and how these are introduced into the system. A number of methods have been proposed to evaluate the vulnerability of karst aquifers, mostly based on GIS methodologies. Some of these methods have been developed for porous and fissured aquifers, such as DRASTIC, AVI, SINTACS, and PI. Other methods were specifically designed for karst aquifer vulnerability mapping: EPIK, COP, the Slovene approach, and PaPRIKa. The use of multiple methods in a specific karst area commonly results in different vulnerability maps, in which reliability is strongly dependent upon quantity, quality, and interpretation of the available data. Despite these differences, these maps should be used as a basis for protection zoning and land-use planning, searching for an acceptable compromise between water protection and economic interests.

Alongside the specific vulnerability assessments, the complexity of karst environments requires a more holistic approach to comprehensively assess the multiple impacts on these fragile geoecosystems. For this purpose, van Beynen and Townsend (2005) proposed the Karst Disturbance Index (KDI), aimed at appraising the impact of multiple human activities and natural processes on the karst environment. The original KDI covers five categories (geomorphology, atmosphere, hydrology, biota, and cultural factors), each one composed of several attributes, in turn subdivided into a number of indicators. This method has satisfactorily been applied in Italy, USA and Jamaica. In the original KDI the evaluator gives numerical scores to each indicator (0 for no disturbance, 3 for almost complete destruction or catastrophic impact). For some indicators, the lack of historical data makes scoring impossible and the assessor has to introduce a "lack of data" (LD), identifying areas in which further research is needed to be able to assign a score. The final value of KDI, comprised between 0 and 1, is obtained by dividing the summatory of the scores of the indicators by the highest possible score. This normalization of the KDI reduces the subjective nature of the assessor's evaluation. A feasible way of simplifying the KDI is to evaluate the disturbances instead of the indicators. Disturbances are similar to indicators, but less specific and thus more readily quantifiable. This modified KDI, which also makes use of normalized values, has successfully been applied in Sardinia, Italy.

A further development of the KDI is the creation of the Karst Sustainability Index (KSI) that takes into account 25 indicators related to the environment, economy, and society.

Source:

Gutiérrez F, et al.(2014). A review on natural and human-induced geohazards and impacts in karst. Earth-Science Reviews. https://doi.org/10.1016/j.earscirev.2014.08.002.

Unit 5   Sandy Desertification

5.1   Conception

The term desertification was apparently coined by the French ecologist LeHouerou (1977) to characterize what was perceived to be a northward advance of the Sahara in Tunisia and Algeria. It gained currency following the severe drought that afflicted the Sud region of Africa in the early 1970s, and again in the 1980s, during which the Sahara was reported to be advancing southward into the Sahelian zone as well. As defined in dictionaries, desertification is the process by which an area becomes (or is made to become) desert-like. The word "desert" itself is derived from the Latin desertus, being the past participle of deserere, meaning to desert, to abandon. The clear implication is that a desert is an area too barren and desolate to support human life. An area that was not originally desert may come to resemble a desert if it loses so much of its formerly usable resources that it can no longer provide adequate subsistence to humans.

Desertification has been neatly defined in the text of the United Nations Convention to Combat Desertification (UNCCD) as "land degradation in arid, semi-arid and dry sub-humid regions resulting from various factors, including climatic variations and human activities".

In recent years, the very term desertification has been called into question as being too vague, and the processes it purports to describe too ill-defined. Some critics have even suggested abandoning the term, in favor of what they consider to be a more precisely definable term, namely, "land degradation". However, desertification has already entered into such common usage that it can no longer be recalled or ignored. It must therefore be clarified and qualified so that its usage may be less ambiguous.

5.2   Areas affected

Drylands currently cover about 46% of the global land area and are home to 300×10^8

people. The range and intensity of desertification have increased in some dryland areas over the past several decades. Desertification hotspots, as identified by a decline in vegetation productivity between the 1980s and 2000s, extended to about 9% of drylands, affecting about 5×10^8 people in 2015. According to the United Nations Convention to Combat Desertification as many as 1.35×10^8 people may be displaced by desertification by 2045, making it one of the most severe environmental challenges facing humanity.

Africa is the continent most affected by desertification, and one of the most obvious natural borders on the landmass is the southern edge of the Sahara Desert. The countries that lie on the edge of the Sahara are among the poorest in the world, and they are subject to periodic droughts that devastate their peoples. African drylands (which include the Sahara, the Kalahari, and the grasslands of East Africa) span 2000×10^4 km^2, some 65% of the continent. 1/3 of Africa's drylands are largely uninhabited arid deserts, while the remaining 2/3 support 2/3 of the continent's burgeoning human population. As Africa's population increases, the productivity of the land supporting this population declines. Some 1/5 of the irrigated cropland, 3/5 of the rain-fed cropland, and 3/4 of the rangeland have been at least moderately harmed by desertification.

Desertified areas in Northern China span dry sub-humid, semi-arid, arid, and hyper-arid climatic 116 zones from northeast to northwest. The desertified area includes five land use types which are sandy 117 land, Gobi, saline land, bare land, and bare rock boulder. In Northern China, the area of the five land 118 use types accounts for 29.3% of the total land area.

5.3 Causes and consequences

The causes of desertification are related to a great amount of factors. For example, the increase of population, the extraction of wood, the excessive cultivation and so on. In summary, inappropriate human activities have led the spread of desertification.

5.3.1 Vegetation destruction

The immediate cause of desertification is the loss of most vegetation. This is driven by a number of factors, alone or in combination, such as drought, climatic shifts, tillage for agriculture, overgrazing and deforestation for fuel or construction materials. Vegetation plays a major role in determining the biological composition of the soil. Studies have shown that, in many environments, the rate of erosion and runoff decreases exponentially with increased vegetation cover. Unprotected, dry soil surfaces blow away with the wind or are washed away by flash floods, leaving infertile lower soil layers that bake in the sun and become an unproductive hardpan.

5.3.2 Climate change

One of the major causes of desertification in the 20th-21st centuries is probably climate change.

In general, desertification is caused by variations in climate and by unsustainable land-management practices in dryland environments. By their very nature, arid and semiarid ecosystems are characterized by sparse or variable rainfall. Thus, climatic changes such as those that result in extended droughts can rapidly reduce the biological productivity of those ecosystems. Such changes may be temporary, lasting only a season, or they may persist over many years and decades. Desertification does a great harm to the climate and it may cause a negative circle. The places with desertification mostly are lack of water, this situation can lead to increased vulnerability of ecosystems. And then, small climates changes can cause the ecological environment here to collapse. As a result, this new desertified areas make the climate there worse.

5.3.3 Overgrazing

Many scientists think that one of the most common causes is overgrazing. There is a suggestion that the last time that the Sahara was converted from savanna to desert it was partially due to overgrazing by the cattle of the local population, too much consumption of vegetation by cattle. Scientists agree that the existence of a desert in the place where the Sahara Desert is now located is due to a natural climate cycle; this cycle often causes a lack of water in the area from time to time. Controversially, Allan Savory has claimed that the controlled movement of herds of livestock, mimicking herds of grazing wildlife, can reverse desertification. Most importantly, the desertification influents our society development and the history of human. Ecological environment is the base of society development. The better the ecology is, the greater the carrying capacity of the population will be, the brighter the prospect of development will be, too.

5.3.4 Irrigated croplands

Nearly 275×10^4 km^2 of croplands are irrigated. Over 60% of these irrigated areas occur in drylands. Certainly, some dryland areas have been irrigated for millennia, but other areas are more fragile. Of the irrigated dryland, 30% (an area roughly the size of Japan) is moderately to severely degraded, and this percentage is increasing.

The main cause of declining biological productivity in irrigated croplands is the accumulation of salts in the soil. There is an important difference between rainwater and the water used for dryland irrigation. Rainwater results from the condensation of water evaporated by sunlight. Essentially, rainwater is distilled seawater or lake water. In contrast, water used for irrigation is the result of runoff from precipitation. Runoff percolates through the soil, dissolving and collecting much of the salts it encounters, before finding its way into rivers or aquifers. When used to irrigate crops, runoff evaporates and leaves behind much of the salts that it collected. Irrigated crops need an average of 80 cm (about 30 inches) of water annually. These salts can build up in the soil unless additional water is used to flush them out. This process can rapidly transform productive land into relatively barren salt flats scattered with halophytes (plants adapted to high levels of salt in the soil).

Most salt-degraded cropland occurs in Asia and southwestern of North America, which account for 75% and 15% of the worldwide total, respectively. In Asia, Iraq has lost over 70% of its irrigated land to salt accumulation. In Russia, much of the irrigated land located where the Volga River runs into the Caspian Sea may last only until the middle of the 21st century before the buildup of salts makes it virtually unusable. Such losses are not restricted to developing countries. In the United States, salt accumulation has lowered crop yields across more than 5×10^4 km², an area that is about a quarter of the country's irrigated land.

5.4 Countermeasures and prevention

Desertification is recognized as a major threat to biodiversity. Some countries have developed Biodiversity Action Plans to counter its effects, particularly in relation to the protection of endangered flora and fauna.

Following a 1977 desertification conference sponsored by the United Nations, the Chinese government took a more proactive approach to prevent and control desertification. Five important programs for forestry as well as desertification control were successively initiated and include the Three North Shelterbelt Development Program (3NSDP, from 1978 to 2050), the Conversion of Cropland to Forest Program (CCFP, since 1999), the Natural Forest Protection Program (NFPP, since 2000), the Sand Source Control Program in the vicinity of Beijing and Tianjin (SSCP, since 2001), and the Grazing Ban for Grassland Restoration Program (GBGRP, since 2003). One of the major objectives of these programs is to mitigate overcultivation, overgrazing, and overcutting, which are the main factors triggering the desertification processes in China.

5.4.1 Reforestation

Reforestation gets at one of the root causes of desertification and is not just a treatment of the symptoms. Environmental organizations work in places where deforestation and desertification are contributing to extreme poverty. There they focus primarily on educating the local population about the dangers of deforestation and sometimes employ them to grow seedlings, which they transfer to severely deforested areas during the rainy season. The Food and Agriculture Organization of the United Nations launched the FAO Drylands Restoration Initiative in 2012 to draw together knowledge and experience on dryland restoration. In 2015, FAO published global guidelines for the restoration of degraded forests and landscapes in drylands, in collaboration with the Turkish Ministry of Forestry and Water Affairs and the Turkish Cooperation and Coordination Agency.

The Green Wall of China is a high-profile example of one method that has been finding success in this battle with desertification. This wall is a much larger-scale version of what

American farmers did in the 1930s to stop the great Midwest dust bowl. This plan was proposed in the late 1970s, and has become a major ecological engineering project that is not predicted to end until the year 2055. According to Chinese reports, there have been nearly 6600×10^8 trees planted in China's great green wall. Due to the success that China has been finding in stopping the spread of desertification, plans are currently be made in Africa to start a "wall" along the borders of the Sahara Desert as well to be financed by the United Nation's Global Environment Facility trust.

5.4.2 Enriching of the soil and restoration fertility

Techniques focus on two aspects: provisioning of water, and fixation and fertilizing soil. Fixating the soil is often done through the use of shelter belts, woodlots and windbreaks. Windbreaks are made from trees and bushes and are used to reduce soil erosion and evapotranspiration. They were widely encouraged by development agencies from the middle of the 1980s in the Sahel area of Africa.

Enriching of the soil and restoration of its fertility is often done by plants. Of these, leguminous plants which extract nitrogen from the air and fix it in the soil, and food crops/trees as grains, barley, beans and dates are the most important. Some research centra are also experimenting with the inoculation of tree species with mycorrhiza in arid zones. The mycorrhiza are basically fungi attaching themselves to the roots of the plants. They hereby create a symbiotic relation with the trees, increasing the surface area of the tree's roots greatly (allowing the tree to gather much more nutrients from the soil).

5.4.3 Farmer-managed natural regeneration

Farmer-managed natural regeneration (FMNR) is another technique that has produced successful results for desert reclamation. Since 1980, this method to reforest degraded landscape has been applied with some success in Niger. This simple and low-cost method has enabled farmers to regenerate some 3×10^4 km^2 in Niger. The process involves enabling native sprouting tree growth through selective pruning of shrub shoots. The residue from pruned trees can be used to provide mulching for fields thus increasing soil water retention and reducing evaporation. Additionally, properly spaced and pruned trees can increase crop yields. The Humbo Assisted Regeneration Project which uses FMNR techniques in Ethiopia has received money from The World Bank's BioCarbon Fund, which supports projects that sequester or conserve carbon in forests or agricultural ecosystems.

5.4.4 Carbon trading and carbon sequestration

The struggle against desertification can occur at several levels. Since regional variations in climate are the main causes of the loss of dryland productivity, it is important to understand the influence of global warming in specific dryland regions. According to some models of climate change, many grasslands in western of North America, for example, are predicted to be at greater risk of drought due to projected increases in summer temperatures and changes to existing rainfall patterns. Many authorities argue that since desertification and global warming are so closely related, one of the main solutions to the former may be the implementation of effective economic policies (such as carbon trading) and technical measures (such as carbon sequestration) that reduce the production of greenhouse gases.

5.4.5 Sustainable land and soil management

Desertification control is a challenge for the sustainable management of land. Climate change and human activities are the two key contributors to desertification and its reversal. Despite the complex interaction of climate change and human factors on desertification, a series of human proactive responses to desertification controlled to a negative growth of desertified land. The strategy and actions of combating desertification include legislation guarantees, policies, science and technology support, national programs, green industry, multisource investments, public participation, and international cooperation.

At local scales, however, desertification is often the result of unsustainable land and soil management. To maintain the biological productivity of the land, soil conservation is often the priority. A number of innovative solutions have been devised that range from relatively simple changes in how people grow crops to labour-intensive landscape engineering projects. Some of the techniques that may help ameliorate the consequences of desertification in irrigated croplands, rain-fed croplands, grazing lands, and dry woodlands include:

①Salt traps, which involve the creation of so-called void layers of gravel and sand at certain depths in the soil. Salt traps prevent salts from reaching the surface of the soil and also help to inhibit water loss.

②Irrigation improvements, which can inhibit water loss from evaporation and prevent salt accumulation. This technique involves changes in the design of irrigation systems to prevent water from pooling or evaporating easily from the soil.

③Cover crops, which prevent soil erosion from wind and water. They can also reduce the local effects of drought. On larger scales, plant cover can help maintain normal rainfall patterns. Cover crops may be perennials or fast-growing annuals.

④Crop rotation, which involves the alternation of different crops on the same plot of land over different growing seasons. This technique can help maintain the productivity of the soil by replenishing critical nutrients removed during harvesting.

⑤Rotational grazing, which is the process of limiting the grazing pressure of livestock in a given area. Livestock are frequently moved to new grazing areas before they cause permanent damage to the plants and soil of any one area.

⑥Terracing, which involves the creation of multiple levels of flat ground that appear as long steps cut into hillsides. The technique slows the pace of runoff, which reduces soil erosion and retards overall water loss.

⑦Contour bunding (or contour bundling), which involves the placement of lines of stones along the natural rises of a landscape, and contour farming. These techniques help to capture and hold rainfall before it can become runoff. They also inhibit wind erosion by keeping the soil heavy and moist.

⑧Windbreaks, which involve the establishment of lines of fast-growing trees planted at right angles to the prevailing surface winds. They are primarily used to slow wind-driven soil erosion but may be used to inhibit the encroachment of sand dunes.

⑨Dune stabilization, which involves the conservation of the plant community living along the sides of dunes. The upper parts of plants help protect the soil from surface winds, whereas the root network below keeps the soil together (Fig. 5-1).

Fig. 5-1 Desertification under control

⑩Charcoal conversion improvements, which include the use of steel or mud kilns or high-pressure compacting equipment to press the wood and other plant residues into briquettes. Conversion improvements retain a greater fraction of the heating potential of fuelwood.

Desertification is really a terrible evolution process. Today, first thing we need to do is changing human's improper activities gradually and then make plans for future. Overall, we human need to join our hands to face the threat (Fig. 5-2).

Fig. 5-2 Green action

Frequently asked questions

(1) How does climate change affect desertification?

Desertification is land degradation in drylands. Climate change and desertification have strong interactions. Desertification affects climate change through loss of fertile soil and vegetation. Soils contain large amounts of carbon, some of which could be released to the atmosphere due to desertification, with important repercussions for the global climate system. The impacts of climate change on desertification are complex and knowledge on the subject is still insufficient. On the one hand, some dryland regions will receive less rainfall and increases in temperatures can reduce soil moisture, harming plant growth. On the other hand, the increase of CO_2 in the atmosphere can enhance plant growth if there are enough water and soil nutrients available.

(2) How can climate change induced desertification be avoided, reduced or reversed?

Managing land sustainably can help avoid, reduce or reverse desertification, and contribute to climate change mitigation and adaptation. Such sustainable land management practices include reducing soil tillage and maintaining plant residues to keep soils covered, planting trees on degraded lands, growing a wider variety of crops, applying efficient irrigation methods, improving rangeland grazing by livestock and many others.

(3) How do sustainable land management practices affect ecosystem services and biodiversity?

Sustainable land management practices help improve ecosystems services and protect biodiversity. For example, conservation agriculture and better rangeland management can increase the production of food and fibres. Planting trees on degraded lands can improve soil fertility and fix carbon in soils. Sustainable land management practices also support biodiversity through habitat protection. Biodiversity protection allows for the safeguarding of precious genetic resources, thus contributing to human well-being.

Source:

IPCC. Climate change and land: an IPCC special report on climate change, desertification, land degradation, sustainable land management, food security, and greenhouse gas fluxes in terrestrial ecosystems (SRCCL) [M/OL]. Cambridge: Cambridge University Press, 2019[2019-08-08]. https://www.ipcc.ch/srccl/

Unit 6 Rocky Desertification

6.1 Conception

Rocky desertification is used to characterize the processes that transform a karst area covered by vegetation and soil into a rocky landscape almost devoid of soil and vegetation (Jiang et al., 2014; Yuan, 1997). The concept of rocky desertification encompasses the physical and ecological changes that take place in the affected land. It involves the loss of vegetation, soil erosion, depletion of soil nutrients, and the exposure of rocky surfaces. These changes are driven by a combination of natural factors, such as climate variations, geological processes, and weathering, as well as human activities like deforestation, overgrazing, improper agricultural practices, and unsustainable land use.

It occurs when surface soil is lost, resulting in bare bedrock, decreased agricultural land value, and degraded ecological environments due to soil erosion. This phenomenon is most common in limestone regions where soil layers are thin, usually less than 10 cm, and the surface appears as a gradual process of rock exposure similar to desert landscapes. In the fragile karst environment with developed carbonate rocks, the original continuous vegetation is destroyed, leading to severe soil loss and exposing large areas of bedrock (Fig. 6-1 and Fig. 6-2).

6.2 Causes

The formation of rocky desertification is primarily driven by intense human activities and is closely related to natural, economic, and social aspects. This results in a combined effect of human factors with natural, environmental, ecological, and geological backgrounds. Rocky desertification has both natural and sociological attributes. The natural attributes of rocky desertification include specific geological backgrounds and processes, biological processes, landscape features, spatial range, and temporal scales. These attributes can be summarized as

Fig. 6-1 Rocky desertification in Guanling County, Guizhou Province of Southwest China

Fig. 6-2 The soils in the region of rocky desertification

different degrees of degradation, occurrence times, levels of energy effects in space, and spatiotemporal manifestations. The sociological attributes of rocky desertification include interactions between humans and the environment, correlation with poverty, and limited controllability. The natural factors that contribute to the formation of rocky desertification mainly include rock solubility, karst processes, topography and landforms, climate conditions, vegetation growth environments, and soil characteristics. On the other hand, human factors include unreasonable land use, reckless deforestation, excessive grazing, and engineering and mining construction.

6.2.1 Natural processes

The abundant carbonate rocks in karst areas are the material basis for the formation of rocky desertification. The steep slopes, warm climate, abundant and concentrated rainfall provide erosion power and dissolution conditions for the formation of rocky desertification. Rocky desertification land formed by natural factors accounts for 26% of the total area of rocky desertification land. The reasons are mainly divided into the following aspects.

①The carbonate rock system has the characteristics of easy leaching and slow soil formation, strong wind erosion resistance, and slow soil formation process.

②The steep surface structure of the mountain is not conducive to the conservation of water and soil resources.

③The special soil profile structure in karst area intensifies soil erosion and rocky desertification on slopes. There is usually a lack of transition layer in the soil profile of karst mountainous areas, and there is a significantly different interface between the matrix carbonate parent rock and the upper soil, which greatly reduces the adhesion and affinity between the rock and soils, and easily leads to soil erosion and rocky desertification when it is stimulated by rainfall.

6.2.2 Human factor

Human factors are the main reasons for the formation of rocky desertification land. Karst areas have high population density, poor regional economy, weak ecological awareness of the masses, and frequent development activities of various unreasonable land resources, resulting in land rocky desertification. The rocky desertification land formed by human factors accounts for 74% of the total area of rocky desertification land.

The reasons are mainly divided into the following aspects.

①Unreasonable farming methods. In karst areas, there are many mountains and few flat lands, and most agricultural production follows the traditional methods of slash-and-burn farming, steep slope farming, wide planting and thin harvest. Due to the lack of necessary water and soil conservation measures and scientific farming methods, abundant and concentrated precipitation makes the soil easily to be eroded, resulting in land rocky desertification.

②Over-cultivation. There is little cultivated land in karst areas. In order to ensure enough cultivated land and solve the problem of food and clothing, the local people often destroy forests and grass for reclamation to expand the cultivated land area and increase the grain production. These newly reclaimed lands, due to the lack of water and soil conservation measures, have serious soil erosion, which finally leads to the disappearance of vegetation and aggravation of rocky desertification.

③Deforestation. Large-scale deforestation of forest resources has led to a significant reduction in forest area and serious damage to forest resources. Due to the loss of surface protection, the development of rocky desertification is accelerated.

④Over-harvesting. In karst areas, the economy is underdeveloped, rural energy types are few, and people's living energy mainly depends on firewood, especially in some areas where coal is short of electricity and energy types are single, firewood mining is the main cause of vegetation destruction.

⑤Grazing indiscriminately. Free-range livestock in karst areas not only destroy forest and grass vegetation, but also cause soil erosion.

⑥The population is growing too fast, the agricultural population is large, and the land load is under pressure.

6.3 Distributions

Karst topography is widely distributed on the earth's surface, covering approximately 2200×10^4 km² of land. This accounts for about 15% of the earth's land surface area. It is inhabited by approximately 100×10^8 people, mainly concentrated in low latitude areas such as Southeast Asia, Southwest China, Central Asia, the Mediterranean, Southern Europe, the Caribbean, the eastern coast of North America, the western coast of South America, and fringe areas of Australia. Concentrated and contiguous karst topography can be mainly found in central and Southern Europe, the eastern part of North America, and the southwestern region of China.

In our country, karst landforms are widely distributed, not only concentrated in the Southwest region but also developing in areas such as North China, Northeast China, Mongolia and Xinjiang, and the Qinghai-Tibet Plateau. However, the largest and most typical karst landforms can be found in the southwest. The karst region of Southwest China includes the Yungui Plateau, Xianggui Hills, Qinghai-Tibet Plateau, and is centered around the Yungui Plateau. It encompasses eight provinces, municipalities, and autonomous regions including Guizhou, Yunnan, Guangxi, Hunan, Hubei, Chongqing, Sichuan, and Guangdong. The exposed area of carbonate rocks exceeds 50×10^4 km², making it the largest and most intensively developed typical region in the world with a concentrated distribution of karst landforms.

The distribution of rocky desertification in our country exhibits the following characteristics: Firstly, it is relatively concentrated, with 81 counties centered around the Yungui Plateau. These counties encompass a mere 27.1% of the karst areas, yet the rocky desertification area within them accounts for 53.4% of the total land area. Secondly, rocky desertification predominantly occurs on steep slopes, with 1100×10^4 hm² of rocky desertification found on slopes above 16°. This steep slope rocky desertification area constitutes 84.9% of the total rocky desertification area. Lastly, mild and moderate rocky desertification types dominate, with mild and moderate rocky desertification land making up 73.2% of the overall rocky desertification area.

6.4 Environmental, social and economic hazard

Rocky desertification can be directly linked to a series of ecological and environmental problems, such as the aggravation of regional soil and water loss, the degradation of ecosystem functions, and the deterioration of rivers, lakes, and reservoirs. Additionally, one of the greatest consequences is the large area of bedrock exposure resulting in a large reduction in cultivating land. The phenomenon of "rocky desertification" after the continuous loss of water and soil resources not only worsens the agricultural production conditions and ecological environment, but also makes the masses lose their basic conditions for survival. Many places have to consider "ecological migration". Rocky desertification areas are prone to flash floods, landslides, debris flows, coupled with underground karst development, resulting in frequent floods and droughts, accompanied by droughts and floods for almost years; At the same time, rocky desertification mountainous areas have high rock exposure rate, little soil, low water storage capacity and strong water leakage, which easily leads to water shortage and drought, and heavy rain will lead to serious soil erosion. Due to serious soil erosion, most areas are short of soil, and some places still have engineering water shortage. Rocky desertification and serious soil erosion have formed a vicious circle, resulting in poor mountains, dry water, declining forests and thin soil, which has given a red light to the survival of local people.

The deteriorating fragile ecological environment in rocky desertification areas restricts regional development, and the population, survival and energy problems in rocky desertification areas have become unavoidable problems.

6.4.1 Environmental hazards

(1) Loss of vegetation

Rocky desertification leads to the degradation and loss of vegetation cover, resulting in reduced biodiversity and habitat loss for native plant and animal species.

(2) Soil erosion

It accelerates soil erosion as the absence of vegetation cover leaves the soil exposed to wind and water erosion, leading to the loss of fertile topsoil and decreased soil quality.

(3) Water scarcity

Rocky desertification reduces the water retention capacity of the landscape, increasing the vulnerability to droughts and water scarcity in affected areas.

(4) Land degradation

The process of rocky desertification contributes to overall land degradation, making it difficult to sustain agriculture, forestry, and other land-based activities.

(5) Diminished ecosystem services

As rocky desertification progresses, it undermines essential ecosystem services such as water filtration, carbon sequestration, and regulation of local climates.

6.4.2 Social hazards

(1) Displacement and migration

Rocky desertification can force rural communities to relocate as the degraded land becomes unsuitable for agriculture or livelihoods, leading to population displacement and migration.

(2) Reduced livelihood opportunities

Loss of arable land and water scarcity can significantly impact the livelihoods of communities reliant on agriculture, livestock herding, and related activities.

(3) Increased poverty

The degradation of land and loss of livelihood opportunities can lead to increased poverty levels in affected areas, impacting the well-being of local communities.

(4) Social inequality

Rocky desertification tends to disproportionately affect marginalized and vulnerable communities, exacerbating social inequalities and worsening the divide between rural and urban populations.

6.4.3 Economic hazards

(1) Decreased agricultural productivity

Rocky desertification hampers agricultural productivity, reducing crop yields and harvests, which negatively impact food security and economic stability.

(2) Increased production costs

Efforts to combat and mitigate rocky desertification often require additional investments in land restoration and management practices, leading to increased production costs for farmers and land managers.

(3) Reduced economic opportunities

The degradation of land limits opportunities for economic activities such as tourism, forestry, and mining, contributing to decreased economic growth and development in affected regions.

Extending reading: What Is Karst?

The term "karst" meaning stony ground, originates in the Dinaric Plateau in the Balkans region of Eastern Europe. This is also where the field of karst science began. Although he may not have been the very first scholar to study karst, Jovan Cviji ć was probably the scientist who laid the foundation for our modern understanding of geomorphic processes in the late nineteenth century. Since then, karst science has expanded to include other types of karst such as relict, paleokarst, pseudokarst, volcanokarst, fluviokarst, and thermokarst. Now in the 21st century, the science of karst has greatly advanced, incorporating an improved understanding of karst environments, their fragility and their value to human development.

Karst is a landscape created by the dissolution of carbonate rocks, although similar features can also be found in volcanic and permafrost areas. Water and its involvement in the process of dissolution is the most significant factor in the creation of karst. It is also of great importance for karst aquifers which are rapidly becoming the most significant issue for karst management. Surface features characteristic of karst include poljes, sinkholes (dolines), swallow holes, karren, pavement of various scales, and dry and blind valleys. Subsurface karst is most commonly thought of by the general public as caves. However, many of these voids cannot be entered by humans as they have no entrances, and it is through these voids or conduits that groundwater can flow. In fact, the presence of these conduits makes karst aquifers difficult to study due to their high degree of heterogeneity with respect to flow rates within the bedrock. Karst can be found around the world, with large regions in Europe, Asia, North and Central America, and the Caribbean.

In the karst areas of Southwestern China the geological environment is extremely fragile, the area is overpopulated and the economy is backward. This has led to serious land degradation in the form of karst rocky desertification, resulting in extensive exposure of the basement rocks. Carbonate rocks cover about 42.6×10^4 km^2, largely in Guizhou Province (11×10^4 km^2), Guangxi Zhuang Autonomous Region (8.9×10^4 km^2) and Yunnan Province (6.1×10^4 km^2). Geomorphically, these karst areas form the transitional terrace between the mountainous western plateau of China and the hilly plain of Eastern China. The land slopes from west to east, from Yunnan (2000~2400 m or more above sea-level) to central Guizhou (1000~1400 m above sea-level) and western Hunan (500~800 m above sea-level). The areas have a population of 1×10^8 and minority nationalities account for about 2000×10^4. There are 48 different nationalities residing in these areas, and they are the most poverty-stricken in China, accounting for half the country's total poverty.

Soil erosion has been a severe ecological environmental concern worldwide, threatening the sustainable development of agriculture and water resources. In particular, soil erosion in karst regions with 10%~15% of the earth's continental surface might lead to more severe environmental problems with rocky land desertification. Under karstification, bedrock fissures develop, leading to karst hillslopes simultaneously exhibiting hydrological conditions of bedrock with and without fissures in karst regions. As a result, unique surface-underground hydrological structures, which are connected by bedrock fissures, are formed due to the unique geological background and hydrochemical conditions, which result in complex surface soil erosion (SSE) and belowground soil erosion (BSE) processes. In some studies, BSE is also defined as soil leakage.

Source:

Fang Q, Zhao L S, Fan C H, et al. How do belowground bedrock fissures impact soil erosion on karst hillslopes under different rainfall intensity conditions? Land Degradation & Development, 2023, 34, 1830-1841.

Peng T, Wang S J. Effects of land use, land cover and rainfall regimes on the surface runoff and soil loss on karst slopes in Southwest China. Catena, 2012, 90, 53-62.

Van Beynen P E. Karst Management. Springer, 2011.

Wang S J, Liu Q M, Zhang D F. Karst rocky desertification in Southwestern China: Geomorphology, landuse, impact and rehabilitation. Land Degradation & Development, 2004(15), 115-121.

Unit 7 Dust Storm

7.1 Conception

A dust storm is a meteorological phenomenon common in arid and semi-arid regions. Dust storms arise when a gust front or other strong wind blows loose sand and dirt from a dry surface. Fine particles are transported by saltation and suspension, a process that moves soil from one place and deposits it in another.

Drylands around North Africa and the Arabian peninsula are the main terrestrial sources of airborne dust. It has been argued that poor management of the earth's drylands, such as neglecting the fallow system, is increasing dust storm's size and frequency from desert margins and changing both the local and global climate, and also impacting local economies.

The term sandstorm is used most often in the context of desert dust storms, especially in the Sahara Desert, or places where sand is a more prevalent soil type than dirt or rock, when a sadstorm occurs, in addition to fine particles obscuring visibility, a considerable amount of larger sand particles are blown closer to the surface. The term dust storm is more likely to be used when finer particles are blown long distances, especially when the dust storm affects urban areas.

7.2 Causes

As the force of wind passing over loosely held particles increases, particles of sand first start to vibrate, then to saltate (leaps). As they repeatedly strike the ground, they loosen and break off smaller particles of dust which then begin to travel in suspension. At wind speeds above that which causes the smallest to suspend, there will be a population of dust grains moving by a range of mechanisms: suspension, saltation and creep.

A study from 2008 finds that the initial saltation of sand particles induces a static electric field by friction. Saltating sand acquires a negative charge relative to the ground which in turn loosens more sand particles which then begin saltating. This process has been found to double the number of particles predicted by previous theories.

Particles become loosely held mainly due to a prolonged drought or arid conditions, and high wind speeds. Gust fronts may be produced by the outflow of rain-cooled air from an intense thunderstorm. Or, the wind gusts may be produced by a dry cold front, that is, a cold front that is moving into a dry air mass and is producing no precipitation—the type of dust storm which was common during the Dust Bowl years in the U. S. Following the passage of a dry cold front, convective instability resulting from cooler air riding overheated ground can maintain the dust storm initiated at the front.

In desert areas, dust and sand storms are most commonly caused by either thunderstorm outflows, or by strong pressure gradients which cause an increase in wind velocity over a wide area. The vertical extent of the dust or sand that is raised is largely determined by the stability of the atmosphere above the ground as well as by the weight of the particulates. In some cases, dust and sand may be confined to a relatively shallow layer by a low-lying temperature inversion. In other instances, dust (but not sand) may be lifted as high as 6100 m (20 000 feet) high.

Drought and wind contribute to the emergence of dust storms, as do poor farming and grazing practices by exposing the dust and sand to the wind. One poor farming practice which contributes to dust storms is dryland farming. Particularly poor dryland farming techniques are intensive tillage or not having established crops or cover crops when storms strike at particularly vulnerable times prior to revegetation. In a semi-arid climate, these practices increase susceptibility to dust storms. However, soil conservation practices may be implemented to control wind erosion.

7.3 Environmental and physical effects

A sandstorm can transport and carry large volumes of sand unexpectedly. Dust storms can carry large amounts of dust, with the leading edge being composed of a wall of thick dust as much as 1.6 km (0.99 mile) high. Dust and sand storms which come off the Sahara Desert are locally known as simoom. The haboob is a sandstorm prevalent in the region of Sudan around Khartoum, with occurrences being most common in the summer.

The Sahara Desert is a key source of dust storms, particularly the Bodélé Depression and an area covering the confluence of Mauritania, Mali, and Algeria. Sahara dust is frequently emitted into the Mediterranean atmosphere and transported by the winds sometimes as far north as central Europe and Great Britain.

Saharan dust storms have increased approximately 10-fold during the half-century since the

1950s, causing topsoil loss in Niger, Chad, Northern Nigeria, and Burkina Faso. In Mauritania, there were just two dust storms a year in the early 1960s, but there are about 80 a year today, according to Andrew Goudie, a professor of geography at Oxford University. Levels of Saharan dust coming off the east coast of Africa in June 2007 were five times those observed in June 2006, and were the highest observed since at least 1999, which may have cooled Atlantic waters enough to slightly reduce hurricane activity in late 2007.

Dust storms cause soil loss from the dry lands, and worse, they preferentially remove organic matter and the nutrient-rich lightest particles, thereby reducing agricultural productivity. Also, the abrasive effect of the storm damages young crop plants. Dust storms also reduced visibility affecting aircraft and road transportation. In addition, dust storms also create problems due to complications of breathing in dust.

Dust can also have beneficial effects where it deposits: Central and South American rain forests get most of their mineral nutrients from the Sahara; iron-poor ocean regions get iron; and dust in Hawaii increases plantain growth. In Northern China as well as the Mid-Western U. S., ancient dust storm deposits known as loess are highly fertile soils, but they are also a significant source of contemporary dust storms when soil-securing vegetation is disturbed.

Dust storms have also been shown to increase the spread of disease across the globe. Virus spores in the ground are blown into the atmosphere by the storms with the minute particles and interact with urban air pollution.

Short-term effects of exposure to desert dust include immediate increased symptoms and worsening of the lung function in individuals with asthma, increased mortality and morbidity from long-transported dust from both Saharan and Asian dust storms suggesting that long-transported dust storm particles adversely affect the circulatory system. Dust pneumonia is the result of large amounts of dust being inhaled.

Prolonged and unprotected exposure of the respiratory system in a dust storm can also cause silicosis, which, if left untreated, will lead to asphyxiation; silicosis is an incurable condition that may also lead to lung cancer. There is also the danger of keratoconjunctivitis sicca (dry eyes) which, in severe cases without immediate and proper treatment, can lead to blindness.

7.4　Source areas

Not all drylands are equally active from a dust storm perspective. Analyses of data from terrestrial meteorological stations and a number of satellite-borne sensors have provided us with a decent, though not perfect, understanding of where the planet's major contemporary desert dust sources are located, although these surveys omit most high latitude sources. The Sahara is undoubtedly the largest source of atmospheric desert dust, contributing perhaps 50% of the global total, followed by China and Central Asia (about 20% of the global total), Arabia and Australia. Drylands in Southern Africa and the Americas are relatively minor sources, together accounting for probably less than 5% of the global total. These regional proportions are by necessity tentative because quantifying the global mineral dust mass budget is not a straightforward matter. Since actual measurements are geographically sparse and temporally sporadic, the majority of estimates are produced using models. Most estimates of total mineral dust emissions to the global atmosphere range $100 \times 10^8 \sim 300 \times 10^8$ t/a. The lack of agreement between models is largely a reflection of the fact that there remains much to learn about dust emission and transport processes

Within dryland regions, certain types of geomorphology are typically richer in dust-sized material than others. Water plays a particularly important role in providing dust-producing desert surfaces such as large basins of internal drainage (e. g. Bodélé, Taoudenni, Seistan, Eyre), alluvial deposits, playas and piedmont alluvial fans. Dust is deflated from these sources with a marked seasonality. The Bodélé Depression produces most of its dust during a season lasting from October to April, affecting much of West Africa with the Harmattan dust haze. In the Middle East, Southern Iraq, Kuwait, Southern Iran and the eastern coast of the Arabian Peninsula are affected by the Shamal, a northwesterly dust-bearing wind that typically prevails between February and October.

Many dust sources are naturally active (e. g. the world's largest source of desert dust-the Sahara's Bodélé Depression) but elsewhere dust is entrained from environments that have become susceptible to wind action through human influence. These situations include lake beds that have been desiccated due to society's use of water, as in the Aral Sea in Central Asia, and agricultural fields left bare after harvests and/or ploughing, as during the notorious Dust Bowl years on the Great Plains of the USA. The mismanagement of rangeland is another frequently cited cause, but the issue of overgrazing is controversial and there are many uncertainties in our understanding of how grasslands may act as dust sources.

Anthropogenic activities may also affect dust emissions indirectly, by changing climate and the hydrological cycle. Climate change is an important potential driver of future wind erosion and dust storm occurrence, especially via the occurrence of more extreme wind events, greater drought frequency and greater aridity in some areas. The relative contribution to global dust emissions of sources significantly influenced by human action is a subject for debate with estimates varying from 50% to insignificant. This high degree of uncertainty can be attributed in large part to the relative lack of detailed characterization of global dust sources.

7.5 Global health impacts of dust storms

7.5.1 Short-term health effects

The short-term effects included all-cause mortality, emergency dispatch or air medical retrieval service, hospitalization or admission, healthcare visits, daily symptoms, decreased pulmonary function, and other problems.

(1) Mortality

Previous researchers have discussed mortality due to dust storms by means of different health problems, such as increased total non-accidental deaths, cardiovascular deaths, mortality due to acute coronary syndrome (ACS), and respiratory mortality. Some studies reported, however, that the number of cases was not increased significantly for all-causes, respiratory, cardiovascular, or cerebrovascular mortality. In Nicosia, the associations for respiratory mortality was −0.79 (−4.69, 3.28) on dust storm days. In Taipei, dust storms have a protective effect on non-accidental deaths, respiratory deaths, and death in people>65 years of age.

(2) Emergency dispatch or air medical retrieval service

Previous researchers discussed the emergency medical services required due to dust storm, focusing on different health problems. This review observed an increased relative risk of all medical emergency dispatches and a significant increase in cardiovascular dispatches, increased daily ambulance calls due to respiratory, cardiovascular, and all causes, and an increase in emergency dispatches due to cardiovascular, respiratory, injury and all causes.

(3) Hospitalization or admission

Hospitalization or admission due to dust storms by means of different health problems or diseases has been discussed in many studies. The results indicated that dust storms were associated with an increased risk of hospital admission due to cardiovascular, cerebrovascular, and respiratory diseases, among others.

Cardiovascular disease (CVD) hospitalizations or admissions. In relation to cardiovascular diseases and the effect of dust storms, 17 studies stated that dust storms can increase: ①The risk of circulatory outpatients and inpatients; ②odds ratio of admission and hospitalization due to congestive cardiac failure and acute coronary syndrome; ③effects on cardiac diseases; ④risk of CVD hospitalization or admission; ⑤emergency admissions for CVD; ⑥the impacts on acute myocardial infarction (AMI); ⑦emergency hospital admissions for ischemic heart diseases (IHD); ⑧hospital admissions for congestive heart failure (CHF); ⑨inpatient hospitalization due to cardiac failure. However, some studies reported non-significant results, such as no association between dust storms and out-of-hospital cardiac arrests and no significant changes in admissions concerning cardiovascular syndromes. Also, some reported no significant association between increased dust particles and angina. It has been reported that the dust storms were not associated with an excess of CVD hospitalizations.

Respiratory disease hospitalizations or admissions. Regarding respiratory diseases related to dust storms, 35 studies stated that dust storms can increase the risk of respiratory outpatients, respiratory disease hospitalizations or admissions, cases of bronchial asthma, asthma-related hospitalizations or admissions, cases of aggravated asthma disease, daily pneumonia admissions, hospital admissions for chronic obstructive pulmonary disease (COPD), emergency hospital admissions for COPD, emergency admissions for respiratory diseases, admitted patients suffering from respiratory infection, and the prevalence of chronic bronchitis, cough, and rhinitis.

Surprisingly, several studies did not find any link between dust storms and negative health outcomes, such as no significant effect on asthma exacerbations in Riyadh, no significant change in the risk of emergency admission in dust events, and no association between sandstorms and risk of hospital admission for asthma or pneumonia patients. Moreover, some studies reported no statistically significant relationship between increased dust levels and pulmonary function, allergic disease, emergency admission, or drug use; no significant relationship between increased risk of chronic obstructive pulmonary disease, asthma, and angina and increased concentration of dust storms; And no excess risk of respiratory hospitalizations. Only two studies found a decrease in respiratory problems after dust storms, like a decreased risk of respiratory inpatients in Taklimakan Desert, and a lower rate of respiration problems among children in areas with higher levels of dust deposition.

Cerebrovascular diseases hospitalizations or admissions. Regarding the correlation between cerebrovascular diseases and dust storms, 6 studies stated that dust storms can increase the risk of cerebrovascular diseases, the incidence of athero-throm-botic brain infarction, stroke admission rates, hospital admissions for epilepsy problems, cerebral ischemic attacks, and various types of headaches, and daily intracerebral hemorrhagic (ICH) stroke admissions. However, previous study reported that sandstorms have no significant relationship with the risk of admission to cerebrovascular patients. Moreover, some studies stated that there is no significant association between the risk of ischemic stroke and dust storms.

Other diseases hospitalizations or admissions. The risk of digestion outpatients and inpatients, gynecology outpatients, pediatrics outpatients and inpatients, and ENT outpatients and inpatients was increased during dust storms. The dust storms were significantly associated with diabetes admissions for females. Furthermore, dust storms can increase the risk of conjunctivitis.

(4) Healthcare visits

Several researchers studied the daily number of healthcare visits due to dust storms for different health problems. Except for 1 article, all others reported that dust storms are associated with an increased daily number of healthcare visits due to asthma-related health problems cardiac, respiratory, and stroke diagnoses, emergency healthcare visits for IHD, CVD, and COPD, conjunctivitis clinic visits, children clinic visits for respiratory problems, healthcare visits for respiratory diseases, healthcare visits for all causes, circulatory, and respiratory diseases, and for cardiovascular and respiratory problems. Moreover, a large increase in emergency visits related to dyspnea during dust storms; however, no clinically significant increase was observed in the total number of emergency visits.

(5) Daily symptoms

Previous studies showed the effects of Kosa on cough and demonstrated that the scores for symptoms (nasopharyngeal, ocular, respiratory, and skin) were significantly higher when related to dust storms. All symptoms (nasal, ocular, respiratory, throat, and skin) increased after exposure to dust storms. An increased risk of eye lacrimation occurrence is associated with dust events. A positive correlation between the increased concentration of dust storms and ocular, nasal, and skin symptoms. Similarly, sandstorms can increase complaints of sleep and psychological disturbances as well as other problems like eye irritation, cough, wheeze, headache, and runny nose.

(6) Pulmonary function

Researchers discussed pulmonary function in relation to dust storms, and the evidence is conflicting. Some studies found that dust storms have a significant, negative effect on pulmonary function. Other studies found no significant relationship between pulmonary function and dust storms. Dust storms can increase the risk of allergic symptoms in pregnant women. A significant increase in respiratory symptoms during dust storms. Besides, sand and dust storms are significantly associated with respiratory symptoms. A relationship between nighttime symptoms and particular matter levels during dust storms. Some studies stated that dust storms worsen respiratory symptoms in asthmatic patients, but some studies reported that pulmonary function was better in children who were more exposed to dust storms than in those with low exposure to dust.

(7) Other impacts

Some experiments explored the relationship of dust storms with road traffic accidents, risk of suicide, placental abruption, and health-related quality of life. Sand storms and the number of vehicles were significantly responsible for road traffic accidents. Dust storms can have adverse effects on the quality of life of patients with asthma and allergies. Dust storms also can decrease health-related quality of life in everyone exposed to them. The exposure to dust storms was associated with an increased risk of suicide (13.1%, $p = 0.002$).

7.5.2 Long-term health effects

Previous studies discussed the long-term adverse health effects caused by dust storms by means of different outcomes, like reduced birth weight, baby's birth weight <2.5 kg, gestation/gestational age >37 weeks and premature birth, and decreased cognitive function in children. Preterm births were correlated with valley fever incidences and increased spring measles incidence. Only one article was observed to indicate no significant effect of desert dust storms on pregnancy consequences.

7.6 Control of dust storms

Dust storms are essentially natural phenomena, which have existed throughout earth's geological ages and on other planets such as Mars. Atmospheric circulation and the existence of deserts on earth determine dust storm sources or origins. However, those storms caused or aggravated by anthropogenic desertification can be eliminated and mitigated. The substantial decrease in dust storms occurring in the United States and the former USSR resulted from a series of adjustment in agricultural practices.

The Chinese government has invested vast social, economic, and scientific resources to suppress dust storms over the past 50 years. A long-term project, covering a broad area from the northeast and north to Northwest China called Three-North Shelter-Forests, started in 1978 to lessen desertification. Although it was successful in some local areas, it has generally been unsuccessful in the arid and semiarid steppe landscape where planted trees cannot survive. Besides this large national endeavor to combat desertification, there were many efforts directed at halting desert encroachment and reclaiming desertified land at a local scale. Strategies included the use of windbreaks and straw checkerboards, land enclosure, irrigating with silt-laden river water, redistribution of material from palaeosols, and chemical treatment. The achievement of the Shapotou Experimental Station in protecting 40 km of the Lanzhou to Baotou railway line from sand burial by the Tengger Desert has been praised. Despite this, the desertification situation at present is not optimistic. It can be summed up by "locally improved, but wholly deteriorated."

A range of sand stabilization and control engineering projects are currently initiated for the purpose of decreasing dust storms in North China. However, combating desertification is problematic. On the one hand, environmental consciousness and ecological sustain? ability are being ignored due to a focus on short-term economic returns. On the other hand, systematic and integrated understanding of environmental deterioration at landscape and regional scales on which sound environmental decisions should be based, are not available. With respect to this dilemma, two actions, which could help diminish the problem, are proposed.

First, the policy of merely focusing onthe production of cropping agriculture should be shifted to ecologically and economically integrated production systems. The environment's long-term value must be recognized. As environmental quality declines, further reductions in agricultural and animal grazing productivity will occur. It is an ethical obligation to prevent environmental degradation and to ameliorate environmental damage. Global failure to control desertification demonstrates that national governments must be convinced to prioritize environmental sustainability.

Second, it is critical at this moment to set a research initiative with a long-term view, focusing on relevant biophysical and social science problems in North China. The available knowledge about desertification is discrete but not yet accepted by land use/management policy-makers. The urgency and importance of the problem cannot justify bypassing scientific procedures. We advocate that a research agenda be proposed as soon as possible. It should link biophysical with socioeconomic issues and monitor problems at various scales in order to develop appropriate desertification control methods for different landscapes and social systems.

Extending reading: Techniques for dust identification and monitoring

The identification of the dust aerosol sources has been a difficult process, due to the complex natural and anthropogenic processes, which are involved in entraining soil particles into the atmosphere during a dust transport. Monitoring of these particles is only possible from satellites because ground-based measurements are very limited in space and time. On the other hand, as desert dust particles get transported over long distances from their source region, they alter their optical properties mainly because of the changes in the number and size of the particles still suspended in the air and the mixing processes in the atmosphere. Therefore, it is important to identify, also for prognostic purposes, the atmospheric circulation patterns facilitating the transport of dust particles from their source regions thousands of kilometers downwind.

In this context, the AI has been proposed as a powerful tool in determining the sources of dust aerosols. The AI is a measure of the wavelength-dependent change in Rayleigh-scattered irradiance from aerosol absorption and is especially suitable for detecting the presence of absorbing aerosols above high reflecting surfaces, such as deserts and snow/ice areas. The absorbing AI, which is linearly proportional to AOD, is defined as the difference between the measured (includes aerosol effects) spectral contrast at the 360~331 nm wavelength radiances and the contrast calculated from the radiative transfer theory for a pure molecular (Rayleigh) atmosphere. AI is mathematically defined as:

$$AI = -100[\lg(I_{360}/I_{331})_{meas} - \lg(I_{360}/I_{331})_{calc}] \tag{7-1}$$

where I is the radiance. Since I_{360} calculation uses reflectivity derived from the I_{331}

measurements, the *AI* definition essentially simplifies to:

$$AI = 100 \lg (I_{360\text{-meas}} / I_{360\text{-calc}}) \qquad (7\text{-}2)$$

The *AI* detects dust, smoke and volcanic ash over all terrestrial surfaces including deserts and snow-ice covered surfaces. These aerosol types are also detected intermingled with clouds and above cloud decks. This is the great advantage between *AI* and *AOD* derived from other satellite sensors, like MODIS. The *AI* can differentiate very well the absorbing and non-absorbing aerosols, by providing a measure of absorption of the UV radiation by smoke and desert dust. The method for the *AI* retrieval via Eq. (7-1) is based on the principle that for a fixed 360 nm radiance the I_{331}/I_{360} spectral contrast is larger for non-absorbing aerosols and clouds and decreases with increasing absorption. Thus, UV absorbing aerosols produce smaller contrast than predicted by the pure Rayleigh scattering atmospheric model; consequently they yield positive values. On the other hand, the non-absorbing aerosols produce greater contrast and negative values. Positive values of *AI* are associated with UV-absorbing aerosols, mainly mineral dust, smoke and volcanic ash. In contrast, negative values are associated with non-absorbing aerosols (e. g. sea-salt and sulfate particles) from both natural and anthropogenic sources. Nevertheless, negative *AI* can be caused by features other than non-absorbing aerosols. Amongst these features are elevated clouds and spectral slopes in the surface albedo between the two wavelengths used for the *AI* detection.

Source:

Kaskaoutis D G, Kambezidis H D, Badarinath K V S, et al. Dust storm identification via satellite remote sensing. 2010.

Unit 8　Ecological Engineering

8.1　Overview

Ecological engineering uses ecology and engineering to predict, design, construct or restore, and manage ecosystems that integrate human society with its natural environment for the benefit of both.

Ecological engineering was introduced by Howard Odum and others as utilizing natural energy sources as the predominant input to manipulate and control environmental systems. The origins of ecological engineering are in Odum's work with ecological modeling and ecosystem simulation to capture holistic macro-patterns of energy and material flows affecting the efficient use of resources. Mitsch and Jorgensen summarized five basic concepts that differentiate ecological engineering from other approaches to addressing problems to benefit society and nature: ①It is based on the self-designing capacity of ecosystems; ②It can be the field (or acid) test of ecological theories; ③It relies on system approaches; ④It conserves non-renewable energy sources; ⑤It supports ecosystem and biological conservation. Mitsch and Jorgensen were the first to define ecological engineering as designing societal services such that they benefit society and nature, and later noted the design should be systems-based, sustainable, and integrate society with its natural environment. Bergen et al. defined ecological engineering as: Utilizing ecological science and theory; applying to all types of ecosystems; adapting engineering design methods; and acknowledging a guiding value system. Barrett (1999) offers a more literal definition of the term: "The design, construction, operation and management (that is, engineering) of landscape/aquatic structures and associated plant and animal communities (that is, ecosystems) to benefit humanity and, often, nature." Barrett continues: "Other terms with equivalent or similar meanings include ecotechnology and two terms most often used in the erosion control field: Soil bioengineering and biotechnical engineering. However, ecological engineering should not be confused with 'biotechnology' when describing genetic engineering at the cellular level, or 'bioengineering' meaning construction of artificial body parts."

The applications in ecological engineering can be classified into 3 spatial scales: ①Mesocosms(0.1 to hundreds of meters); ②Ecosystems (one to tens of kilometers); ③Regional systems (>tens of kilometers). The complexity of the design likely increases with the spatial scale. Applications are increasing in breadth and depth, and likely impacting the field's definition, as more opportunities to design and use ecosystems as interfaces between society and nature are explored. Implementation of ecological engineering has focused on the creation or restoration of ecosystems, from degraded wetlands to multi-celled tubs and greenhouses that integrate microbial, fish, and plant services to process human wastewater into products such as fertilizers, flowers, and drinking water. Applications of ecological engineering in cities have emerged from collaboration with other fields such as landscape architecture, urban planning, and urban horticulture, to address human health and biodiversity, as targeted by the UN Sustainable Development Goals, with holistic projects such as stormwater management. Applications of ecological engineering in rural landscapes have included wetland treatment and community reforestation through traditional ecological knowledge.

Ecological engineering utilizessystems ecology with engineering design to obtain a holistic view of the interactions within and between society and nature. Ecosystem simulation with Energy Systems Language (also known as energy circuit language or energese) by Howard Odum is one illustration of this systems ecology approach. This holistic model development and simulation defines the system of interest, identifies the system's boundary, and diagrams how energy and material moves into, within, and out of, a system in order to identify how to use renewable resources through ecosystem processes and increase sustainability. The system it describes is a collection (i.e.group) of components (i.e.parts), connected by some type of interaction or interrelationship, that collectively responds to some stimulus or demand and fulfills some specific purpose or function. By understanding systems ecology the ecological engineer can more efficiently design with ecosystem components and processes within the design, utilize renewable energy and resources, and increase sustainability.

8.2 Terrace

Agriculture terrace is a traditional farming method which involves the construction of flat or sloping fields on steep slopes, hillsides, or mountains. The fields are designed with a series of raised terraces or steps, usually made of stone or soil, which allows farmers to cultivate crops on steep land surfaces without soil erosion. The agriculture terrace originated thousands of years ago and is still in use today in various parts of the world. It has been found to be particularly useful in areas with limited arable land, where farmers need to maximize their yield by cultivating every available space.

As a widespread and famous type of farming system, terraced agriculture offers numerous benefits. The terraces help to reduce soil erosion by reducing the speed of water flow and preventing soil movement down the steep slopes. The terraces can also be designed to capture rainfall, prevent surface runoff, and recharge aquifers, ensuring that water is conserved and used efficiently. Terraced agriculture can result in higher yields by providing a better growing environment for crops, such as improving soil fertility and drainage. Additionally, terraced agriculture allows for the cultivation of crops in areas that would otherwise be difficult or impossible to farm. Overall, agriculture terracing is an important farming practice that helps to sustainably support food production in steep, mountainous terrain while minimizing the impact of agriculture on fragile ecosystems.

Terraces are substantially an efficient ecological engineering of soil & water conservation. Terraces decrease the length of the hillside slope, reducing rill erosion. By diverting or attenuating overland flow, they prevent the formation of gullies, allow sediment to settle from runoff water, and improve the quality of runoff water leaving the field. In drier areas, terraces serve to retain runoff and increase the amount of water available for crop production. Such retention of water also reduces the risk of wind erosion. Terracing can aid in surface irrigation on steeper land, particularly in paddy rice production.

In the practice of construction, terrace alignment can either be non-parallel or parallel. Non-parallel terraces follow the contour of the land regardless of alignment. Some minor adjustments are frequently made to eliminate sharp turns and short rows by installing additional outlets, using variable grade, and installing vegetated turning strips. Non-parallel terraces are best suited to applications other than row crop farming, such as small grain agriculture or pastures. Parallel terraces are preferred for row crop farming operations. They generally require greater cut and fill volumes during construction than non-parallel systems.

The design of a terrace system includes specifying the proper spacing and location of terraces, the design of a channel with adequate capacity, and the development of a stable and sometimes farmable cross section. For the graded terrace, runoff must be removed at nonerosive velocities in both the channel and the outlet. Soil characteristics, cropping and soil management practices, and climatic conditions are the most important considerations in terrace design.

Proper maintenance is as important as the original construction of the terrace. However, it need not be expensive since normal farming operations will usually suffice. Any breakovers should be repaired as soon as possible. The terrace should be watched more carefully during the first year after construction, and any excessive settlement, failures, or cracking repaired. Channels may occasionally need to be cleared of deposited sediment or ridges rebuilt.

8.3 Conservation tillage

Conservation tillage refers to a series of practices that involves minimal disturbance of the soil in order to conserve its natural structure, fertility, and organic matter. This involves reducing or eliminating tillage, combining crop residue retention, and rotating crops. The primary benefit of conservation tillage is that it can help to reduce soil erosion and improve soil health, leading to increased crop productivity and sustainability. It also has the potential to reduce water pollution, greenhouse gas emissions, and energy use, making it an environmentally friendly farming method. However, conservation tillage can also represent a challenge for farmers, as it requires careful management practices and may lead to increased weed pressure, nutrient management issues, and pest problems.

No-till never uses a plow, disk, etc. ever again, and aims for 100% ground cover. No-till farming (also called zero tillage or direct drilling) is a way of growing crops or pasture from year to year without disturbing the soil through tillage. No-till is an agricultural technique that increases the amount of water that infiltrates into the soil, the soil's retention of organic matter and its cycling of nutrients. In many agricultural regions, it can reduce or eliminate soil erosion. It increases the amount and variety of life in and on the soil, including disease-causing organisms and disease organisms. The most powerful benefit of no-tillage is improvement in soil biological fertility, making soils more resilient. Farm operations are made much more efficient, particularly improved time of sowing and better trafficability of farm operations. No-till farming improves soil quality (soil function), carbon, organic matter, aggregates, protecting from erosion, evaporation of water, and structural breakdown. Reducing of tillage reduces compaction of soil. This can help reduce soil erosion almost to soil production rates.

Strip-till is a conservation system that uses a minimum tillage. Narrow strips are tilled where seeds will be planted, leaving the soil in between the rows untilled. It combines the soil drying and warming benefits of conventional tillage with the soil-protecting advantages of no-till by disturbing only the portion of the soil that is to contain the seed row. This type of tillage is performed with special equipment and can require the farmer to make multiple trips, depending on the strip-till implement used, and field conditions. Each row that has been strip-tilled is usually about eight to ten inches wide.

Strip till warms the soil, it allows an aerobic condition, and it allows for a better seedbed than no-till. Strip-till allows the soil's nutrients to be better adapted to the plant's needs, while still giving residue cover to the soil between the rows. The system will still allow for some soil water contact that could cause erosion, however, the amount of erosion on a strip-tilled field would be light compared to the amount of erosion on an intensively tilled field. Furthermore, when liquid fertilizer is being applied, it can be directly applied in these rows where the seed is being planted, reducing the amount of fertilizer needed while improving proximity of the fertilizer to the roots. Compared to intensive tillage, strip tillage saves considerable time and money. Strip tillage can reduce the amount of trips through a field down to two or possibly one trip when using a strip till implement combined with other machinery such as a planter, fertilizer spreader, and chemical sprayer. This can save the farmer a considerable amount of time and fuel, while reducing soil compaction due to few passes in a field. With the use of GPS-guided tractors, this precision farming can increase overall yields. Strip-till conserves more soil moisture compared to intensive tillage systems. However, compared to no-till, strip-till may in some cases reduce soil moisture

In agriculture mulch tillage or mulch-till fall under the umbrella term of conservation tillage and refer to seeding methods where a hundred percent of the soil surface is disturbed by tillage whereby crop residues are mixed with the soil and a certain amount of residues remain on the soil surface. A great variety of cultivator implements are used to perform mulch-till. Mulch is material to regulate heat. This is done by covering it with any material like wood chips, straw, leaves or food waste.

8.4 Windbreak

A windbreak (shelterbelt) is one or more rows of trees or shrubs planted to provide protection from wind, prevent soil erosion, and reduce water loss. Shelterbelts are commonly planted in hedgerows around the edges of fields in agricultural areas to protect crops, livestock, and homes from the harsh effects of wind. Shelterbelts can also provide habitats for wildlife and improve air quality by mitigating dust and pollution. Other benefits include contributing to a microclimate around crops with slightly less drying and chilling at night.

The shelterbelt can be planted in a variety of patterns and densities. Shelterbelts and intercropping can be combined in a farming practice referred to as alley cropping. Fields are planted in rows of different crops surrounded by rows of trees. These trees provide fruit, wood, or protect the crops from the wind. A further use for a shelterbelt is to screen a farm from a main road or motorway. This improves the farm landscape by reducing the visual incursion of the motorway, mitigating noise from the traffic and providing a safe barrier between farm animals and the road.

In essence, when the wind encounters a porous obstacle such as a windbreak or shelterbelt, air pressure increases on the windward side and conversely decreases on the leeward side. As a result, the airstream approaching the barrier is retarded, and a proportion of it is displaced up and over the barrier, resulting in a jet of higher wind speed aloft. The remainder of the impinging airstream, having been retarded in its approach, now circulates through the barrier to its downstream edge, pushed along by the decrease in pressure across the shelterbelt's width; emerging on the downwind side, that airstream is now further retarded by an adverse pressure gradient, because in the lee of the barrier, with increasing downwind distance air pressure recovers again to the ambient level. The result is that minimum wind speed occurs not at or within the windbreak, nor at its downwind edge, but further downwind-nominally, at a distance of about 3~5 times the windbreak height H. Beyond that point wind speed recovers, aided by downward momentum transport from the overlying, faster-moving stream. From the perspective of the Reynolds-averaged Navier-Stokes equations these effects can be understood as resulting from the loss of momentum caused by the drag of leaves and branches and would be represented by the body force (a distributed momentum sink).

Not only is the mean (average) wind speed reduced in the lee of the shelter, the wind is also less gusty, for turbulent wind fluctuations are also damped. As a result, turbulent vertical mixing is weaker in the lee of the barrier than it is upwind, and interesting secondary microclimatic effects result. For instance, by day sensible heat rising from the ground due to the absorption of sunlight (see surface energy budget) is mixed upward less efficiently in the lee of a windbreak, with the result that air temperature near ground is somewhat higher in the lee than on the windward side. Of course, this effect is attenuated with increasing downwind distance and indeed, beyond about $8H$ downstream a zone may exist that is actually cooler than upwind.

8.5 Sand fence

A sand fence or sand break is used to force windblown, drifting sand to accumulate in a desired place. Sand fences are employed to control erosion, help sand dune stabilization, keep sand off roadways, and to recruit new material in desert areas.

A typical construction is to attach aperforated plastic sheet to stakes at regular intervals, similar to construction site fencing or temporary sports field fencing. Another is a cedar or other lightweight wood strip and wire fence, also attached to metal stakes. A permanent sand fence is generally of larger wooden poles set deeply into the ground with large wooden planks running horizontally across them.

The drifting and settling of sand behind and in front of such a fence occurs because the wind speed on both the downwind and windward sides is less than that on the far windward side, allowing light materials such as sand to settle. This creates a pile both in front of and behind the sand fence causing more sand to drop out. Conveniently the sand does not drop on the barrier itself, otherwise it would soon be buried and rendered useless.

Sand fences are appropriate for areas with loose sands that high winds can transport off-site. They can be helpful for construction sites with large areas of cleared land or in arid regions where blowing sand is problematic. Shorefront development sites also benefit because sand fences help to form frontal dunes.

Sand fences are only effective when construction staff place them perpendicular or near perpendicular to the prevailing wind. They should be at least 3~4 feet high with an effective minimum porosity of 50%. For wooden slat fences, this means the gap between slats should equal the slat width. For prefabricated commercial products, such as woven polyethylene, the manufacturer should specify the porosity and the material should be ultraviolet-resistant. Construction staff should also install woven fences with a gap of 1~2 feet between the fabric bottom and ground surface to prevent breakage during high winds. Erecting multiple rows of fences, spaced 20~40 feet apart, increases their overall sediment-trapping efficiency. To protect stockpiles, construction staff should place wind fencing upwind of the stockpiles at a distance of approximately three times the height of the stockpile. In coastal dune areas, construction staff should place sand fences away from the mean high tide line. Using native vegetation can enhance fence integrity. Sand fences do not control sediment carried in stormwater discharges and are not effective for dust control. Where erosion control and dust control are necessary, install sand fences with other erosion and sediment control practices.

Extensive Reading: Ecological slope engineering
(1) Vegetation establishment and succession

Ecological slope engineering measures usually go through three major ecological phases: (I) Vegetation establishment, (II) vegetation succession towards improved stability, and (III) long-term system stability management in order to maintain a functional vegetation cover. For many classic slope-engineering issues, for example stabilization of cut slopes or mine tailings, the initial phase of vegetation establishment is most sensitive to numerous forms of erosion, since stable root systems are yet to develop. Success of planting efforts and growth rate depend on the choice of species, climate, growing conditions, and irregular events, such as weather extremes.

Generally, herbaceous vegetation performs better during the establishment phase as compared to woody vegetation, since plant density of herbaceous species is usually higher and they need less time to build up significant fine-root systems. However, some shrubs can lead to satisfactory results as well. Further, fertilization, mulching, irrigation and nets can increase the success of herbaceous vegetation establishment, albeit such efforts can be costly. Random factors, such as weather extremes, are unpredictable. Nevertheless, if planning efforts account for seasonal weather probabilities, the risk of heavy rain events or droughts during this sensitive initial phase can be minimized.

For vegetation establishment, a biodiverse planting approach may lead to better results. As mentioned above, growing conditions are not homogenous over the slope and a carefully selected diverse seed mixture can improve the plant growth rate over a range of different growing conditions (nutrient and water supply, soil pH value, soil microorganisms, etc.). The choice of local species may also improve slope stabilization performance and vegetation establishment. Finally, competition for soil resources during the establishment phase can lead to dominance of undesirable plant species, which may reduce slope stabilization. Management measures, such as mowing, can prevent such undesirable vegetation shifts. If establishment of woody vegetation is preferred on a slope, the combination with herbaceous species is recommended in order to reduce the probability of slope failure, and of rill-and gully erosion during vegetation establishment.

Over time, root concentration and thus soil reinforcement usually increases until root growth and root decay reach a more or less stable equilibrium. However, temporal root system dynamics can also vary by species. For example, for barley (*Hordeum vulgare* L.), an initial increase of RAR with time was followed by a decrease due to increased root decay. However, this decrease did not seem to affect the measured soil reinforcement. Comino and Druetta observed a similar trend with perennial ryegrass (*Lolium perenne* L.), probably due to changing weather conditions during their experiment. Despite their importance, such temporal changes remain poorly researched.

After the primary phase of stabilization, the vegetation goes through a long succession process towards a denser and ecologically more stable ground cover. In some cases, however, succession can lead to temporal or permanent reduction of slope reinforcement. If high levels of reinforcement are required, such changes should be taken into account during planning. Moreover, changes in management, e.g. overuse or abandonment of pastures, can lead to successional changes towards reduced slope stability. Below the tree line, succession of herbaceous vegetation gradually leads to afforestation, if no countermeasures are taken (e.g. cutting, grazing). Therefore, consistent management of herbaceous vegetated slopes is an important tool for steering the succession of the vegetation community towards a desirable result, for example a stabilizing permanent herbaceous vegetation cover, or gradual afforestation.

Vegetation succession and related management measures should be an integrated part of ecological slope engineering measures. Accounting for dynamic effects of succession at an early stage may help to maintain a consistent functional vegetation cover and reduce long-term costs. This way, the succession can be guided towards a resistant, resilient, and mechanically stable vegetation cover.

(2) The sum of the pieces

Ecological slope engineering is the establishment and management of artificial, or semi-natural ecosystems that foster one or several ecosystem services in our favor. Ecological slope engineering aims at fostering the ecosystem service of slope stabilization. Focus has been mostly on mechanical stability assessment of the current state, or an expected state, while ecological aspects of vegetation communities (i. e. dynamic changes) have been often neglected. Consequently, ecological descriptions of slope engineering measures are rare. Nevertheless, dynamic changes of slope stabilizing vegetation communities can play an important role for short and long-term changes in slope stabilization performance. Such changes can depend on many factors, including management, changing conditions of the physical and chemical surroundings, seasonal dynamics, or natural succession. Profound understanding of such ecological short and long-term interactions plays an important role in the success of ecological slope engineering measures.

In this context, vegetation management should be understood as anthropogenic alteration of environmental conditions that favor the provision of certain ecosystem services, including slope stabilization. In the long-term vegetation management and thus also slope stabilization performance. Furthermore, management can prevent vegetation succession towards unfavorable vegetation communities, for example dominance of species with poor root systems, or afforestation. Therefore, including the dynamics of ecosystems in the planning can reduce the risk of soil damages and costs for related remediation measures, both in the short and long-term.

Coppin and Richards describe bioengineering as "an 'art', based on engineering experience and judgement, rather than an exact 'science'", due to the diverse character and the necessity to adapt management measures to specific local conditions. If designed and managed in the right way, spatial and temporal synergies between plant species can lead to an improved overall ecosystem performance (i. e. biomass production, root concentration, resource use efficiency), which exceeds the sum of the performance of all individual elements. Consequently, slope stabilization performance may improve as well.

(3) Ecological slope engineering practice

As we made clear above, a general strategy for ecological slope engineering with herbaceous vegetation should be to rely on a diversity of preferably local, non-invasive plant species, since they usually grow well in their natural habitats. Diverse natural communities are usually less sensitive to environmental changes. In optimal designs, chosen species should additionally maintain other ecosystem services, in order to create multi-functional, biodiverse landscapes. Additional ecosystem services could include watershed management, water retention and filtration, carbon sequestration, food and fodder production, or the maintenance of an esthetic landscape with high recreational value. In summary, an ideal plant ecosystem for slope engineering should be resilient and resistant, require little maintenance, and should maintain a number of different ecosystem services.

A classic concept for diversification of plant functional traits is the structural diversification in terms of a multi-dimensional vegetation approach. While a one dimensional vegetation cover includes only one vegetation type, a multi-dimensional approach mimics natural systems, where herbaceous vegetation (undergrowth) is often combined with shrubs and trees. Accordingly, different ecological growing strategies and slope stabilization characteristics of species of different vegetation types can be combined. For example deep-reaching root anchors of woody vegetation can be combined with the high topsoil reinforcement potential of herbaceous vegetation. Hence, the soil is stabilized in deeper layers where needed, due to the woody vegetation, while herbaceous vegetation prevents shallow slope failure in the topsoil layer through soil reinforcement and the surfacemat effect, as well as buttressing and arching. Nevertheless, there is little research or experience with assessing, planning, and managing such multi-dimensional slope stabilizing ecosystems. A combination of field studies and advanced computer simulations may allow slope stability assessment of such complex, structurally diversified ecosystems.

Locally, this approach may even allow the combination of ecological slope engineering demands with existing agroforestry concepts.

Source:

Lobmann M T, et al. The influence of herbaceous vegetation on slope stability-A review. Earth-Science Reviews. https://doi.org/10.1016/j.earscirev.2020.103328

Unit 9 Watershed Management

9.1 Overview

Every body of water (e.g. rivers, lakes, ponds, streams, and estuaries) has a watershed. The watershed is the area of land that drains or sheds water into a specific receiving waterbody, such as a lake or a river. As rainwater or melted snow runs downhill in the watershed, it collects and transports sediment and other materials and deposits them into the receiving waterbody. Watershed management is a term used to describe the process of implementing land use practices and water management practices to protect and improve the quality of the water and other natural resources within a watershed by managing the use of those land and water resources in a comprehensive manner. Watershed management aims at the sustainable distribution of its resources and the process of creating and implementing plans, programs, and projects to sustain and enhance watershed functions that affect the plant, animal, and human communities within the watershed boundary.

Runoff from rainwater or snowmelt can contribute significant amounts of pollution into the lake or river. Watershed management helps to control pollution of the water and other natural resources in the watershed by identifying the different kinds of pollution present in the watershed and how those pollutants are transported, and recommending ways to reduce or eliminate those pollution sources.

All activities that occur within a watershed will somehow affect that watershed's natural resources and water quality. New land development, runoff from already-developed areas, agricultural activities, and household activities such as gardening/lawn care, septic system use/maintenance, water diversion and car maintenance all can affect the quality of the resources within a watershed. Watershed management planning comprehensively identifies those activities that affect the health of the watershed and makes recommendations to properly address them so that adverse impacts from pollution are reduced.

Watershed management is also important because the planning process results in a partnership among all affected parties in the watershed. That partnership is essential to the successful management of the land and water resources in the watershed since all partners have a stake in the health of the watershed. It is also an efficient way to prioritize the implementation of watershed management plans in times when resources may be limited.

Because watershed boundaries do not coincide with political boundaries, the actions of adjacent municipalities upstream can have as much of an impact on the downstream municipality's land and water resources as those actions carried out locally. Impacts from upstream sources can sometimes undermine the efforts of downstream municipalities to control pollution. Comprehensive planning for the resources within the entire watershed, with participation and commitment from all municipalities in the watershed, is critical to protecting the health of the watershed's resources.

9.2 Integrated water resources management

Water resources management is the process of managing water resources in order to meet the water needs of people and the environment. It includes managing water supplies, planning and maintaining infrastructure to collect, treat and distribute water, and regulating water quality to ensure that it is safe for drinking and other uses. Effective water resources management requires a comprehensive understanding of water use, availability and quality, along with the knowledge and skills to implement policies and programs that effectively balance the competing demands of different water users and stakeholders. This includes ensuring that water is allocated fairly and efficiently, minimizing water waste and pollution, and addressing the impacts of climate change and other environmental factors on water resources.

Integrated water resources management is defined as a process which promotes the coordinated development and management of water, land and related resources, in order to maximize the resultant economic and social welfare in an equitable manner without compromising the sustainability of vital ecosystems. This concept aims to promote changes in practices which are considered fundamental to improved water resource management.

Integrated water resources management rests upon three principles that together act as the overall framework. Firstly, it ensures equal access for all users (particularly marginalized and poorer user groups) to an adequate quantity and quality of water necessary to sustain human well-being. Secondly, it brings the greatest benefit to the greatest number of users possible with the available financial and water resources. Finally, it requires that aquatic ecosystems are acknowledged as users and that adequate allocation is made to sustain their natural functioning.

Operationally, integrated water resources management approaches involve applying knowledge from various disciplines as well as the insights from diverse stakeholders to devise and implement efficient, equitable and sustainable solutions to water and development problems. As such, integrated water resources management is a comprehensive, participatory planning and implementation tool for managing and developing water resources in a way that balances social and economic needs, and that ensures the protection of ecosystems for future generations.

9.3 Environmental laws

Environmental laws often dictate the planning and actions that agencies take to manage watersheds. Some laws require that planning be done, others can be used to make a plan legally enforceable and others set out the ground rules for what can and cannot be done in development and planning. Most countries and states have their own laws regarding watershed management. The below lists some laws governing watershed stewardship by U. S. Forest Service.

Organic Administration Act of 1897 defines the original purposes of national forests-to improve and protect the forests, to furnish a continuous supply of timber, and to secure favorable conditions of water flows. Years of concern about watershed damage led to creation of the national forests. Watersheds must be cared for to sustain their watershed function as sponge-and-filter systems that store water and naturally regulate runoff. The goals are good plant and ground cover, streams in dynamic balance with their channels and flood plains, and natural conveyance of water and sediment.

Multiple Use-Sustained Yield Act of 1960 amplifies national forest purposes to include watershed, wildlife and fish, outdoor recreation, range, and timber. Renewable surface resources must be managed for multiple use and sustained yield of the several products and services they provide. The principles of multiple use and sustained yield include the provision that land productivity shall not be impaired.

Endangered Species Act of 1973 was written to conserve endangered and threatened species of wildlife, fish, plants, and the ecosystems on which they depend. Federal agencies must conserve endangered & threatened species in cooperation with state and local agencies. Conservation means using all means needed to recover species to where the measures provided pursuant to this law are no longer needed. Each agency shall ensure that actions are unlikely to jeopardize the continued existence of any endangered or threatened species or result in adverse modification of their critical habitat.

National Forest Management Act of 1976 was written to guide forest planning and management. Programs must protect and, where appropriate, improve the quality of soil and water. Timber must be harvested only where soil, slope, and watershed conditions are not irreversibly damaged; the land can be adequately restocked within five years after harvest; and streams, lakes, wetlands, and other water bodies are protected from detrimental impacts.

Federal Land Policy and Management Act of 1976 was written to guide management of national forests and grasslands. These lands must be managed to protect ecological, environmental, air, water resource, and other values, and provide food and habitat for fish, wildlife, and domestic animals. Rights-of-way and uses shall include terms and conditions to protect the environment, subject to valid existing rights.

Clean Water Act of 1977 was written to restore and maintain the chemical, physical, and biological integrity of the nation's waters. The goal is to sustain the integrity of water quality and aquatic habitat so that waters of the United States (perennial and intermittent streams, lakes, wetlands, and their tributaries) will support diverse, productive, stable, aquatic ecosystems with a balanced range of aquatic habitats. The Forest Service must comply with water quality laws and rules like anyone else and must apply best management practices to protect water quality.

Extensive reading: Watershed structures

The concept of *Watershed Management* (WSM) involves planning the development of a resource region. Resource region incorporates the private property as well as common property regimes. The focus of planning is the optimal and sustainable use of the resources viz., water and land. The approach of watershed treatment therefore ought to be holistic and involves quite a few hydrological, biophysical and socioeconomic aspects. WSM being a land based technology, it would help conserve and improve in situ soil moisture, check soil erosion and improve water resources, especially groundwater in the rainfed regions. It simply means improving the management of a watershed or rainfall catchments area through technical and non-technical interventions. Technical interventions are required in order to adopt the hydrological and biophysical conditions to the needs of the local communities. At the same time non-technical (socioeconomic) instruments are required to make the technical interventions more effective.

Large variety of interventions are being adopted in WSM, most of them fall within the field of soil and water conservation. These include both technical and non-technical measures, which in aggregate form help to limit the rate of soil loss and provide sustainable land and water use. Some measures may help prevent erosion whilst others may only control it, thus limiting the extent and timing of its impact. There are also interventions which are aimed at rehabilitating soils after erosion has negatively impacted the land, which include mechanical soil conservation practices; vegetative cover, afforestation, building contour bunds, water harvesting structures (farm ponds, check-dams), field bunds (raised edges), ridge bunds, etc. All these interventions are expected to facilitate higher land productivity through improved overall ecological conditions such as moisture and water availability for agriculture. It may be noted that all these interventions are not adopted in all the watersheds or regions. Nature and intensity of interventions vary according to the hydrogeology and bio-physical conditions of the region or watershed. Here we discuss these interventions in brief. The technical watershed interventions can be grouped under three broad typologies: Soil conservation; in situ moisture conservation; water harvesting structures.

(1) Soil conservation

Majority of soil conservation programs instituted in the1970's and 1980's were dominated by engineered systems. These systems were originally developed for large land holdings in temperate regions. Some of them are adopted in South Asian Sountries. Most popular ones include: ①Lock and spill drains-graded drain which is usually built along contours and acts by capturing runoff in small stilling ponds, which allows the rainwater to infiltrate slowly; ②Stone walls-constructed along contours in fields or road sides and provide an irregular form of terracing; ③Bench terraces-construction of large benches on steep slopes and cultivation may be carried out on slightly downward or back sloping surfaces; ④Bunds-artificial embankments constructed and graded so as to intercept rainfall and sediments and lead runoff away from the cultivated land.

These interventions are widely used across the countries though they are mostly implemented on the common lands and steep slopes. Besides, these approaches are commonly used in the centrally managed tea estates in Sri Lanka, where land is not in short supply and valuable topsoil can be sacrificed for the construction of bunds and terraces. Pebble bunding is also popular in the regions where soil quality is poor and crop lands are covered with pebbles. Clearing the crop lands of pebbles and using them to make bunds across gullies and small streams observed to serve double purpose of improving the quality of land and checking soil erosion.

Maintaining the vegetative cover is an effective way of reducing runoff and soil erosion. In fact, it has been observed that vegetative cover is more effective in checking oil erosion as well as improving water yields or stream flows. Calder has estimated that afforestation will reduce river flows by the order of 200 mm/a in Sri Lanka. Afforestation is also widely adopted in the watershed management, especially at the ridge locations. Afforestation and forest conservation measures are extensively used in the hilly terrains of Nepal, Bangladesh, India, and Pakistan. Most of these soil conservation interventions are not very popular among the farming communities in South Asia. This is mainly due to small size land holding. These interventions are either capital intensive or land intensive or both. Often governments provide incentives or subsidies to promote on farm soil conservation activities. But they do not sustain once the incentives dry up. One activity that is observed to be very popular is the pebble bunding in many parts of the state of Andhra Pradesh, India. This is mainly due to the nature of the terrain.

(2) In situ moisture conservation

Increasing the water holding capacity of soils in the arid and semi-arid regions is a main concern for enhancing the agricultural productivity. Soil erosion further erodes the water holding capacity in these regions as, organic matter and clays contribute disproportionately more to water-holding capacity relative to other coarser soil fractions. Number of interventions are adopted to increase in situ soil moisture content, which are known as on farm interventions. These on farm activities are most widely promoted and adopted measures in the rainfed regions of India, especially during the initial years of 1970s and 1980s. These include on farm bunding, trenching and contour bunding to increase soil moisture within plots of land by reducing the run off. These interventions help retaining the rainwater within the farmers own field in the arid and semi-arid zones, where annual rainfall is low (200~600 mm). In India, these interventions are implemented with 90%~95% percent government subsidy. These interventions, however, are land intensive and are not favoured by small and marginal farmers. In the case of very small plots, the loss of area offsets the yield gains.

Agronomically, there are a number of farming practices that can help increase soil moisture content and enhance yield rates. These recommended interventions include: vegetative soil cover, mulching, use of simple tillage practices, contour cultivation, ridging across the slope, vegetative barriers, ripped furrows, land levelling, level pans, terracing, etc. These practices, however, are only recommendations and not implemented as part of the watershed programme. Number of research studies have shown that these practices enhanced soil quality and yield.

(3) Water Harvesting Structures

Of late rainwater harvesting is increasingly gaining popularity among local communities as well as policy makers. Often, this specific component draws a substantial share in the watershed allocations. For instance, in India the density and size of rainwater harvesting structures is on the rise during the last two decades. In the low rainfall arid and semi-arid regions, these structures provide a life line to protect crops and livestock during lean months. Rainwater harvesting structures can be constructed at the surface as well as subsurface level, though surface structures are the main interventions in the watershed programmes. The structures include check-dams; farm ponds; percolation tanks or pits; recharge wells and injection wells. All these interventions help to recharge groundwater. While check dams and farm ponds are part of water spreading methods, which serve the purpose of recharge as well as direct use of water;

percolation pits, percolation tanks, recharge wells and injection wells are used for recharging the groundwater.

The adoption of these methods depend on the location (upstream/downstream) within the watershed as well as the hydrogeology of the location. In a ridge to valley context of watersheds, it is often recommended that on-farm interventions (in-situ soil moisture conservation) and water spread methods (farm ponds and check dams) in the middle and downstream locations. However, understanding the hydrogeology is critical for determining the suitability of these interventions. From the hydrogeology perspective, moderate to deep weathering and fracturing zones are suitable for artificial recharge through water spreading methods (farm ponds and check dams); areas with deep fractures are suitable for artificial recharge methods by injection methods depending on the aquifer position; and areas with very shallow basement are not suitable for any interventions. Based on the drainage order, mini-percolation and percolation tanks are effective on the first to third order streams. Check dams are more effective when the topography is plane. The interventions required for watersheds located on the weathered zone are different from watersheds located on the fractured zone. Water-spreading methods such as check dams, percolation tanks, and farm ponds are effective in weathered zones.

The impacts of rainwater harvesting methods are more conspicuous than any other interventions of watershed management. The visible collection of water at the check dams and the impact on the surrounding wells attract local community's attention. As a result, farmers demand more of these structures, as these structures can't be built by individual farmers due to their capital intensity. Besides, check dams needs to be constructed in the natural streams, which are common property. Even the implementing agencies are leaning towards creating more of these structures due to the demands from the communities and also due to their quick and visible impacts. These structures, however, provide location specific impacts and hence are not equitable in terms of distribution of benefits. Moreover, their impacts could be very limited when they are constructed in the absence of hydrogeological information of the watershed.

Source:

Ratna R V, et al. Watershed management in South Asia: A synoptic review. Journal of Hydrology. https://doi.org/10.1016/j.jhydrol.2017.05.043

Chapter 2

Academic English Writing

Unit 10　Characters of Academic English

10.1　Vocabulary characteristics

(1) 具有国际性(internationalism)

从词源学角度看，专业英语词汇多来源于希腊语、拉丁语，因此，比普通英语更具有国际性。据尼贝肯(Oscar E. Nybakken)统计，普通英语词汇中有46%直接或间接来自拉丁语，有7.2%来自希腊语；而在专业英语词汇中，源于希腊语、拉丁语的多达70%以上，如 botany(植物学)、spleen(脾)、electron(电子)、larynx(喉)等。

(2) 使用规范的书面动词(using standard written verb)

在动词方面，专业英语较多地使用规范的书面语动词来代替日常口语中的动词短语。例如，用 discover 来代替 find out，用 observe 代替 look at 等。

(3) 大量使用派生的形容词(extensive use of derived adjectives)

为了描述各种科学现象和物质特性，专业英语中有许多表示数量、大小、程度、性质、状态的形容词。除一般形容词外，大多数由动词、名词派生而来。例如，在动词后加-able、-ed、-ing、-ive 等，在名词后加-al、-ic、-ious 等。

10.2　Features of grammatical structure

(1) 大量使用名词化结构(nominalization)

由于专业文体要求行文简洁、表达客观、内容确切、信息量大、强调存在的事实而非某一行为，在专业英语中，名词的使用率比动词高。

【例句】The exploded drawing is designed to show several parts in their proper locations prior to assembly, so we call it a type of pictorial drawing. /The exploded drawing is a type of pictorial drawing designed to show several parts in their proper locations prior to assembly.

【译文】分解图是在进行装配前以适当的位置显示各个零件的一种视图。

【注释】由于专业英语所表述的是客观规律,尽量避免使用第一、第二人称。此外,应使主要信息置于句首。

【例句】The earth rotates on its own axis, which causes the change from day to night. / The rotation of the earth on its own axis causes the change from day to night.

【译文】地球绕轴自转,引起昼夜的变化。

【注释】名词化结构 The rotation of the earth on its own axis 使复合句简化成简单句,而且使表达的概念更加准确严密。

(2) 广泛使用被动语句

科技文章侧重叙事推理,强调客观准确。第一、二人称使用过多,会造成主观臆断的印象,因此尽量使用第三人称叙述,采用被动语态。

【例句】Desertification is widely recognized as land degradation in arid, semi-arid, and dry sub-damp areas.

【译文】荒漠化被广泛认为是干旱、半干旱和亚湿润干旱地区的土地退化。

【例句】Soil resistance to erosion can be measured as soil erodibility K factor.

【译文】土壤抗侵蚀性可以用土壤可蚀性 K 因子来衡量。

(3) 非谓语动词

专业英语语言要求文章简练、结构紧凑,因此,常使用分词短语代替定语从句或状语从句;使用分词独立结构代替状语从句或并列分句;使用不定式短语代替各种从句;使用"介词+动名词短语"代替定语从句或状语从句。

①分词短语。具有形容词和副词的语法功能,在句中充当定语、状语等,现在分词(present participle)强调动作的主动和正在进行,而过去分词(past participle)用于被动语态和完成时态中,强调动作的被动和完成。

【例句】A soil develops from some starting material such as consolidated rock or unconsolidated material deposited by wind or water. The starting material is the parent material for a soil.

【译文】土壤是由风或水沉积的固结岩石或松散物质等初始物质形成的,起始物质是土壤母质。

【例句】The physical property of soil is an important basis for making management measures such as rational tillage and irrigation and drainage.

【译文】土壤的物理性质是制定合理耕作、合理排灌等管理措施的重要依据。

②动名词。是起到名词作用的一种动词形式,在句中充当主语、宾语、表语、定语等成分。

【例句】The erosion is both downward, deepening the valley, and headward, extending the valley into the hillside, creating head cuts and steep banks.

【译文】侵蚀既向下,加深山谷,也向前,将山谷延伸到山坡上,形成了头部切口和陡峭的河岸。

【例句】Thermal erosion is the result of melting and weakening permafrost due to moving water.

【译文】热侵蚀是由于流动的水融化和削弱永久冻土的结果。

③动词不定式。是用法最多的非谓语动词形式，具有名词、形容词或副词的语法功能。在句中充当主语、宾语、表语、定语、状语等成分。

【例句】Erosion control has the potential to sequester carbon as well as restoring degraded soils and improving water quality.

【译文】控制侵蚀具有固碳以及恢复退化土壤和改善水质的潜力。

【例句】Further increases in precipitation are sufficient to overcome the protective effect.

【译文】降水的进一步增加足以克服保护作用。

(4) 复合词与缩写词

例如，Feed-back（动词+名词构成名词）、Full-enclosed（形容词+分词构成形容词）、VPD(vapor pressure deficit)水蒸气压差、CAD(computer assistance design)计算机辅助设计。

10.3 Features of commonly used sentence

(1) 常用定语从句

根据定语从句与先行词的关系，定语从句可分为限制性定语从句和非限制性定语从句。限制性定语从句紧跟先行词，主句与从句不用逗号分开，从句不可省去。非限制性定语从句与主句之间有逗号分开，起补充说明作用，省去后意思仍完整。

①限制性定语从句。that 指人或物，在从句中作主语时不可省略，作宾语可省略。which 指物，在从句中作主语时不可省略，作宾语可省略。

【例句】The windbreak(which) they constructed last year was effective.

【译文】他们去年修筑的防风林很有效。

当从句中含介词时，只能用"介词+which"结构。

【例句】A line on which the value of each point was the minimum at the corresponding height separates the region A and B.

【译文】A 区和 B 区由一条线隔开，该线上每个点的值在相应高度处为最小值。

Listen to, look at, depend on, pay attention to, take care of 等固定短语动词，在从句中不宜将介词与动词分开。

【例句】This is the issue which the researchers pay attention to.

【译文】这是研究人员所关注的问题。

关系词只能用 that 的 3 种情况：

一是先行词被序数词或形容词最高级所修饰，或本身是序数词、基数词、形容词最高级时。

【例句】The highest value that shows the actual situation are shown in the table 1.

【译文】表1表示了实际情况的最高值。

二是先行词为 all、any、much、many、everything、anything、none、the one 等不定代词时，或被 the only、the very、the same、the last、little、few 等词修饰时。

【例句】This is the only related literature that I can find.

【译文】这是我能找到的唯一文献。

②先行词表示时间、地点、原因时用关系副词 when、where、why。

【例句】Environmental organizations work in places where deforestation and desertification are contributing to extreme poverty.

【译文】环境组织在森林砍伐和荒漠化导致极端贫困的地方工作。

【例句】A further area of high erosion risk occurs where the landforms and associated soils are relics of a previous climate.

【译文】另一个高侵蚀风险的区域发生在以前气候所遗留的地形和相关土壤上。

③Which 引导的非限制性定语从句。是先行词的附加说明，去掉也不影响主句的意思。换页言之，非限制性定语从句是对主句或主句中的主语或宾语进行信息的补充说明，而不是真正意义上的起修饰或限制作用。它与主句之间通常用逗号分开。如果放在句子中间，则前后都需要用逗号隔开。关系代词 which 在非限制性定语从句中所指代和修饰的可以是主句中的名词、形容词、短语、其他从句或整个主句，在从句中作主语、动词宾语、介词宾语或表语。非限制性定语从句的关系词在句子中不能省略。

特殊结构：名词/代词 + of + which/whom/who，例如，light is the fast thing in the world, the speed of which is 300000 km per second。

【例句】The strength of soldered joint is less than a joint which is brazed, riveted or welded.

【译文】锡焊点的连接力要小于钎焊、铆接或焊接的连接点。

(2) 常用 it 作形式主语

【例句】In some cases it may be necessary to connect metal surfaces by means of a hard spelter solder which fuses at hight temperature.

【译文】某些情况下有必要用一种高熔融温度的强锌焊料来连接金属表面。

【注释】在这种结构中，it 本身并不具备实际的意思，只是用来强调客观事实。因此，在翻译时不能刻板地直译为"它"。

【例句】It has been argued that poor management of the earth's drylands, such as neglecting the fallow system, are increasing dust storms size and frequency from desert margins and changing both the local and global climate, and also impacting local economies.

【译文】有人认为，对地球上旱地的管理不善，如忽视休耕系统，正在增加来自沙漠边缘的沙尘暴的规模和频率，改变了当地和全球的气候，也影响了当地的经济。

【例句】It should be stressed that the general trends described above are often masked by the high variability in erosion rates for any given quantity of precipitation as a result of differences in soil, slopes and land cover.

【译文】应当强调的是，上述一般趋势往往被由于土壤、斜坡和土地覆盖的不同而导致的任何特定降水量的侵蚀率的高度变异性所掩盖。

(3) 常用由 as 引导的主动、被动及简略形式

例如，as the illustration shows（如图所示）、as has been stated（如前所述）、as follows（如下）等。

(4) 常用包含两个以上从句的长句

【例句】There is a suggestion that the last time that the Sahara was converted from savanna to desert it was partially due to overgrazing by the cattle of the local population.

【译文】有一种说法认为，上次撒哈拉沙漠从热带草原转变为沙漠的部分原因是当地居民的过度放牧。

【例句】In agriculture, a terrace is a piece of sloped plane that has been cut into a series of successively receding flat surfaces or platforms, which resemble steps, for the purposes of more effective farming.

【译文】在农业中，梯田是一块倾斜的平面，它被切割成一系列连续后退的平面或平台，类似于台阶，以便实现更有效的农业。

【例句】Thus, the concept of receptor implies that the same molecule that has the bending or recognition site can participate normally in the initiating event in cell activation.

【译文】因此，受体（的概念）是具有结合或识别（化学信使）的位点，在正常情况下参与细胞活动启动的分子。

【注释】这句话的主语是 the concept of receptor，其谓语动词是 implies。谓语动词后面是由 that 引导的宾语从句，而在从句的主语后又由 that 引到了一个修饰主语的定语从句。

10.4 Exercises

通过练习，理解下列句式的语法结构和句式特点。

【例句】Upright fences, usually made of wood bars, bunches of straws or reeds, or tree branches, are widely used to check drifting sand and drifting snow because of their easy availability, low cost, and simple construction.

【译文】直立栅栏通常由木条、一捆捆稻草、芦苇或树枝组成，由于其容易获得、成本低和施工简单，被广泛用于控制流沙和飞雪。

【例句】The higher turbulence produced by low-porosity fences may result in the recovery of mean horizontal wind velocities to levels equal to upwind velocities at a distance closer to the fence, thereby decreasing the shelter distance.

【译文】低孔隙度栅栏产生的较高湍流度可能导致平均水平风速恢复到与靠近栅栏距离的逆风速度相等的水平，从而减少遮蔽距离。

【例句】Current approaches to estimate threshold friction velocity (TFV) of soil particle movement, including both experimental and empirical methods, suffer from various disadvantages, and they are particularly not effective to estimate TFVs at regional to global scales.

【译文】目前估算土壤颗粒运动阈值摩擦速度（TFV）的方法，包括实验方法和经验方法，都存在着各种各样的缺点，在区域尺度和全球尺度上对阀值摩擦速度的估算尤其无效。

【例句】Understanding the resistance and resilience of foundation plant species to climate change is a critical issue because the loss of these species would fundamentally reshape communities and ecosystem processes.

【译文】了解基础植物物种对气候变化的抵抗力和恢复力是一个关键问题，因为这些物种的消失将从根本上重塑群落和生态系统进程。

【例句】Grasslands and shrublands provide fundamental ecosystem services in arid and semi-arid regions but are highly susceptible to desertification induced by anthropogenic and climatic drivers.

【译文】草原和灌木地在干旱和半干旱地区提供基本的生态系统服务，但极易受到人为和气候因素引起的沙漠化的影响。

【例句】Scientific planning for soil and water conservation requires knowledge of the relations between those factors that cause loss of soil and water and those that help to reduce such losses.

【译文】科学规划水土保持需要了解造成水土流失的因素与有助于减少水土流失的因素之间的关系。

【例句】Desertification is caused by serious land degradation resulting from the forced use of already marginal land resources and their subsequent abandonment, and in the past few decades has been aggravated by drought.

【译文】沙漠化是由于过度使用已经退化的土地资源而造成的更为严重土地退化，而这些资源随后被遗弃，在过去几十年里，干旱加剧了沙漠化。

【例句】This is a non-circulating blow-type wind tunnel, capable wind speeds ranging from 4~35 m/s, with a measurement precision of ±3%~±0.5%.

【译文】该风洞为非循环吹风式风洞，风速范围为4~35 m/s，测量精度为±3%~±0.5%。

【例句】At each measuring point, greater than 100 instantaneous wind speeds were recorded and the mean values were calculated and denoted as U.

【译文】在每个测点，记录大于 100 的瞬时风速，计算平均值，记为 U。

【例句】An unpolished wooden board with a thickness of 4 cm, serving as a roughness element, was installed on the floor at the upwind boundary of the test section to facilitate the formation of an atmospheric boundary layer.

【译文】在测试段迎风边界的地面上安装一块厚度为 4 cm 的未抛光木板作为粗糙度单元，以促进大气边界层的形成。

【例句】Aeolian landforms are found in areas where wind is the primary agent of transport, such as in arid and semiarid regions, while elsewhere the effects of aeolian processes are often masked by the effects of hydrologic processes.

【译文】风成地貌出现在以风为主要运输媒介的地区，如干旱和半干旱地区，而在其他地方，风成过程的影响往往被水文过程的影响所掩盖。

【例句】The most recognizable evidences of aeolian activity on the earth surface are the sand dunes with different forms and sizes observed in desert and coastal environments.

【译文】地球表面风沙活动最明显的证据是在沙漠和海岸环境中观察到的不同形状和大小的沙丘。

【例句】In a typical dust-related traffic incident on the highway, suspension of dust-sized particles may cause the deterioration of visibility whereas the near-surface transport and deposition of saltation-sized particles may reduce the traction on the road surface.

【译文】在典型的公路粉尘相关交通事故中，粉尘颗粒悬浮可能导致能见度下降，而粉尘颗粒的近地表运输和沉积可能降低路面的牵引力。

【例句】In Northern China, wind is typically the strongest from March to May, corresponding to a time period when the frequency and intensity of aeolian sand transportation is also the strongest. However, at this time of the year, the windbreaks constructed of or dominated by deciduous plants still present the winter facies and thus provide poor protection.

【译文】在中国北方，风的强度一般在 3~5 月最强，对应风沙输送的频率和强度也最强的时间段。然而，在每年的这个时候，由落叶植物构成或占主导地位的防风林仍然呈现冬季相，因此提供的保护很差。

【例句】Russian olive, native to western to central Asia, is a thorny median-size shrub or tree that can be a dominant species in desert riparian ecosystems. It is also an important tree species used for shelterbelt construction in many regions, including Northwestern China, Western North America, and Turkey.

【译文】俄罗斯橄榄原产于中亚西部，是一种多刺的中等大小的灌木或乔木，可能是沙漠沿岸生态系统的优势物种。它也是一个重要的树种，在许多地区，包括中国西北部、北美西部和土耳其，用于防护林带建设。

【例句】Windbreaks have been used for centuries to shelter crops from wind damage and to protect soils from wind erosion. They reduce wind speed and alter the characteristics of airflow around them, inducing changes in the surrounding atmospheric, plant, and soil environments

【译文】几个世纪以来，防风林一直被用来保护农作物免受风害，保护土壤免受风蚀。它们降低风速，改变周围气流的特性，从而引起周围大气、植物和土壤环境的变化。

【例句】The scope of Aeolian Research includes the following topics: Fundamental Aeolian processes, including sand and dust entrainment, transport and deposition of sediment; Modeling and field studies of Aeolian processes; Instrumentation and measurement in the field and lab; Practical applications including environmental impacts and erosion control; Aeolian landforms, geomorphology and paleo environments; Dust-atmosphere/cloud interactions.

【译文】风成研究的范围包括以下主题：基本风成过程，包括沙尘夹带、搬运和沉积物沉积；风成过程的模拟和实地研究；现场和实验室的仪器和测量；实际应用包括环境影响和侵蚀控制；风成地貌、地貌和古环境；尘—气/云的相互作用。

【例句】The growing use of and reduced costs for remote sensing technology in gully erosion research might very well transform soil erosion models and improve natural resource assessments and management efficiency.

【译文】在沟壑侵蚀研究中，遥感技术被广泛应用且成本较低，这也许能很好地改进土壤侵蚀模型以提高自然资源评估和管理效率。

【例句】The system can reduce wind speed by 70% and the sand transportation rate by 96% when wind reached the farmland after passing through the oasis-protection system.

【译文】该系统通过防风系统后，风速可降低70%，风沙输送率可降低96%。

【例句】For an optimal oasis-protection system, the width and water requirements of the vegetation barriers were also crucial factors because water resources are an important limiting factor for maintaining vegetation barriers in the long-term.

【译文】对于最优的绿洲保护系统，植被屏障的宽度和需水量也是至关重要的因素，因为水资源是长期维持植被屏障的重要限制因素。

【例句】High speed winds blow soil from the desert, depositing some on neighboring fertile lands, and causing shifting sand dunes within the desert, which buries fences and blocks roads and railway tracks.

【译文】高速的风将土壤从沙漠吹来，将一些土壤沉积在邻近的肥沃土地上，并在沙漠中造成沙丘移动，从而掩埋了篱笆，堵塞了公路和铁路。

【例句】Generally, aerodynamic porosity and optical porosity are not equivalent, since the latter does not account for the three-dimensional nature of pores through which the wind flows.

【译文】一般来说，空气动力孔隙度和光孔隙度是不相等的，因为光孔隙度不考虑风通过的孔隙的三维性质。

【例句】The extreme nature of drylands means that semi-arid vegetation is often patchy and dynamic through time and space.

【译文】干旱土地的极端性质意味着半干旱的植被往往是零碎的，在时间和空间上是动态的。

【例句】They are also very common in agricultural areas because sustained agriculture depletes the soil of much of its organic content, increasing the erodibility of the soil.

【译文】它们在农业地区也很常见，因为持续的农业消耗了土壤的大部分有机含量，增加了土壤的可蚀性。

【例句】Salt stress alters plant metabolism, reduces the endogenous water potential, and causes ionic imbalance and lipid peroxidation, which altogether lead to considerable inhibition of plant growth.

【译文】盐胁迫会改变植物的代谢，降低内源水势，引起离子失衡和脂质过氧化，对植物生长具有相当大的抑制作用。

【例句】Fences that are always constructed to have optical porosities greater than zero are important artificial windbreaks. They can be classified as upright, horizontal, griddled, holed-plank, and wind-screened, depending on the available materials.

【译文】栅栏的光学孔隙度总是大于零，是重要的人工防风林。根据可用材料的不同，它们可以分为直立式、卧式、格状、孔板和风幕式。

【例句】Windbreak shelter is modeled in terms of friction velocity reduction, which is a function of wind speed and direction, distance from the barrier, windbreak height, porosity, width, and orientation.

【译文】防风林防护效益是根据风速和方向、与屏障的距离、防风林的高度、孔隙度、宽度和方向的函数来建模的。

【例句】On a regional scale, windbreak systems increase terrain roughness, which means that a dense network slows down the average wind speed of the region as a whole.

【译文】在区域尺度上，防风系统增加了地形的粗糙度，这意味着密集的网络降低了整个区域的平均风速。

【例句】Six major windbreak types were identified using Drachenfels (1994) classification, which was slightly modified to meet the characteristics of the windbreaks in the area.

【译文】采用 Drachenfels (1994) 分类确定了 6 种主要的防风林类型，并对其进行了轻微的修改，以满足该地区防风林的特点。

【例句】Some of them are very wide, which fits them to their secondary purpose of wildlife reserves.

【译文】其中一些非常宽,这符合它们作为野生动物保护区的次要用途。

【例句】The core of the model is the erosion module, which simulates soil loss or deposition due to creep, saltation, and suspension transport.

【译文】该模型的核心是侵蚀模块,它模拟了由于蠕变、跃变和悬浮运输引起的土壤流失或沉积。

【例句】These symposia have published 150 peer-reviewed papers, many of which have had a measurable impact in the discipline.

【译文】这些专题论文集发表了150篇同行评议的论文,其中许多对该学科产生了可衡量的影响。

【例句】Rates of soil loss due to permanent gullies significantly exceed losses observed in agricultural areas, which can be the leading cause of landscape degradation worldwide.

【译文】永久性沟壑造成的土壤损失率大大超过农业区的土壤损失率,这可能是全球景观退化的主要原因。

【例句】Topographic indices are typically expressed as a product of local slope and upstream drainage area, which are used as proxies for stream power and the susceptibility for landsliding.

【译文】地形指数通常表示为局部坡度和上游排水面积的乘积,可用来代替水流功率和滑坡敏感性。

【例句】For the grassland, heavy grazing reduces vegetative cover as well as causes severe soil compaction, both of which increase erosion rates.

【译文】对于草地,过度放牧减少了植被覆盖,并造成严重的土壤压实,两者都增加了侵蚀速率。

Unit 11　Writing up Research

　　科技论文(实验研究报告)是研究人员为描述所开展的研究而撰写的论文(报告)。论文(报告)的目的是向该领域的其他研究人员解释研究的目标、实验方法和研究结果，是高年级本科生与研究生进行学术研究的基本技能。论文分为研究性论文和综述性论文两大类。研究性论文是基于新开展的实验研究、实地考察调研而获得数据和结果并加以讨论，获得新知识和原创性的科学研究论文。从内容上可分为研究性论文、技术或方法论文(提出新的或改良的技术或工艺，或介绍新的分析软件、计算方法、工作流程或数据标准)和数据论义(提供科学记录、数据或数据集并介绍数据采集、整理、编目、标准化或评估的过程)等。论文(报告)可以发表在专业期刊，或以专著的形式出现或作为大学学位要求的一部分，还可以硕士论文或博士学位论文的形式撰写。

　　可以开展的研究(实验)有多种类型。例如，研究人员为了识别和控制可能影响某种研究结果产生的尽可能多的因素而进行实际验证时进行的科学控制实验；相关性研究使研究人员有机会比较两个或多个不同的变量，以确定它们之间是否存在任何可预测的关系。其他类型的研究可能涉及从调查问卷或案例中获得的信息，也可以使用计算机生成的模型来进行，这些模型用于解释或预测在实验室或自然界中观察到的现象。

　　所有这些类型的科学研究都有一些共同的特点，即它们都是围绕一个科学问题而设计的。为了提供该科学问题的可能答案，研究人员设置一个假设，然后以推翻或支持该假设的方式设计科学实验。这类科学研究是定量的，通常会用一个或多个统计测试来对所获得的数据进行测试，以确定所获结果的可靠性，使研究者对此做出是否应当对该结果进行重视的决定。因此，撰写这些不同类型的论文(研究报告)也有很多共同点。

　　通常，论文(实验研究的出版物)在结构上由标题(题目)、摘要、引言、材料和方法、结果和讨论组成。摘要包括对研究目的、方法和结果的描述。完整的结果通常以表格和图表的形式在结果部分呈现。一篇论文(报告)还引用了同一研究领域其他已发表论文的信息。参考书目(参考文献列表)以及其他所需要的所有信息都需要在图书馆中查找,这个参考文献列表总会包含在论文(报告)的末尾。最后,投稿时通常还会附上一份涵盖论文(报告)中最重要信息的简要自荐信。

当发表研究论文时,你可能会发现,想要发表的每个期刊的结构要求都不尽相同。每个期刊可能会根据其期刊风格提出不同的要求。例如,*Environmental Geology*、*Ecological Engineering*、*Catena*、*Agriculture Ecosystems & Environment* 的要求都会各不相同。但是,在这些水土保持与荒漠化防治领域常见的期刊上发表研究成果时,需要查阅相应期刊的格式要求。以下介绍科技论文常见的撰写规范和基本要求。

11.1 Formulating a research question

尽管科学问题很少出现在最终的论文(报告)中,但是,科学问题的提出是设计和开展科学研究的基础。只有当研究人员提出特定的研究主题后,为了解决他们感兴趣研究主题中的特定问题,就会首先提出一个科学问题。例如,"What are the effects of increased concentrations of sulfuric acid in the atmosphere on production of grain sorghum?"和"Whether the observed GPP (gross primary productivity)-NDVI (normalized difference vegetation index) relations may hold equally for stressed and non-stressed vegetation?"都是科学问题的例子。

在正式的研究工作中,有必要设定一个研究预期所能产生的结果,这就是所谓的假说。该假说可能是对科学问题的响应。当假设以否定的方式陈述时,它被称为无效假设(H_0)。例如,"We hypothesize that the lower but constant rates of nitrogen addition due to chronic deposition will not lead to a detectable transient increase in foliar nitrogen concentration"就是一个无效的假设。假设也可以是非否定的。例如,"We hypothesized that soil water shortage is dominant over other stresses in controlling net ecosystem carbon exchange of dry land ecosystems"和"We propose that the seasonal dynamics of leaf area index is an important determinant of productivity over the growing season"。实验的目的是决定这一假设是否可以被推翻的前提。如果无效假设被推翻,研究者就得到了预期的实验结果。否则,研究者未能推翻无效假设,将无法获得预期实验结果。

11.2 Title

当拟定研究论文的题目(标题)时,将论文的内容浓缩成几个字,吸引读者、审稿人、编辑的注意力,并将论文与同一主题领域的其他论文区分开来,这一点非常关键。

首先,论文的题目需要保持简洁和具吸引力。论文题目(标题)的主要功能是为读者提供论文准确简洁的研究内容。因此,论文题目(标题)为保持简洁明了,词汇使用时应运用主动动词而不是基于名词的复杂短语，并且应避免不必要的细节。此外，一个好的研究论文题目通常有 10~12 个单词。一个冗长的论文题目（标题）可能看起来没有重点，会令读者的注意力从真正重要的点上转移开。此外，论文题目（标题）应使用适当的描述性词语。一个好的研究论文题目（标题）应该包含初稿（手稿）中使用的关键词，并应定义研究内容的性质。

论文题目不能过大或过于空泛，也不宜过长、过于详细具体。要避免使用"关于……的研究"（Study on ..., Research on ..., An investigation of ..., Experiments of ..., Experimental study of ...）或"对……的思考""关于……的思考"（Thoughts on ..., Regarding...）之类的套语。尽可能避免使用化学式、上下角标、公式、特殊符号（数字符号、希腊字母等）和不必要的专业表达。但是，如果某些术语全称较长且缩写词已得到领域公知公用，是可以在标题中使用的。此外，分子式在化学或医学等领域的文章中有时也不可避免地出现。系列标题意味着论文是系列中的某一篇文章，内容是不完整的，应避免使用。标题中应包含有助于编制索引和计算机搜索的主题词或关键词，但应使标题清晰易懂、具体、信息丰富。

11.3 Abstract

除了论文（报告）其他部分内容提供的信息外，摘要主要为读者提供研究的简要信息预览。几乎所有研究领域论文的摘要内容及格式都是以非常相似的方式撰写。摘要包含的信息类型及其顺序安排非常传统。例如，摘要需要提供一些研究的背景信息、主要研究内容（研究目的）及其研究范围、研究方法、研究最重要的结果以及结论或建议。

在格式方面，一些期刊的摘要限定字数。因此，需要减少（之前撰写的）摘要的字数。精简摘要通常只关注 2~3 个信息要素，摘要的重点应放在研究结果。

在语言方面，撰写摘要时使用的动词时态与论文（报告）前面相应部分中使用的时态直接相关。例如，摘要的背景信息与引言第一阶段的背景信息相似，需使用现在时。"One of the basic principles of communication is that the message should be understood by the intended audience"是一个用现在时态介绍背景信息的例子。"In this study, the readability of tax booklets from nine states was evaluated"是运用过去时介绍背景的例子。"Net energy analyses have been carried out for eight trajectories which convert energy source into heated domestic water"是结合过去时和现在完成时介绍背景的例子。材料与方法和结果部分一般使用过去时态。例如，"Children performed a 5-trial task"和"Older workers surpassed younger ones in both speed and skill jobs"使用的都是过去时态。在撰写结论部分时，可使用现在时、情态动词或语气助词。例如，"The results suggest that the presence of unique sets of industry factors can be used to explain variation in economic growth"是使用情态动词的例子。

摘要范文
范文 1

The quantitative carbon dynamics of desertified lands in Northern China were predicted for the years of 2000 and 2030, based on the areas and conversion rates (positive and negative) of desertified lands in the past forty years and organic carbon contents of soils. The top 1.0 m soil layer of natural desertified lands in China contained some 7,841 Tg of organic carbon as of 1992. In China, over the last 40 years, a total of 2,812 Tg of organic carbon was released from desert lands and, in the reverse process about 644 Tg of organic carbon were fixed into lands undergoing desertification. Thus, China's desert lands have shown a net release of 2168 Tg of organic-C over the past 40 years, equivalent to 7,949 Tg of CO_2. By the year 2000, the area of desertified lands in China had increased 40,300 km^2 and released 368 Tg of organic carbon into the atmosphere. By 2030, this area will increase to 249,700 km^2 and release about 1996 Tg of organic carbon into the atmosphere. Net releases of 151 Tg and 1,243 Tg of organic carbon can be expected by the year of 2000 and 2030, respectively. This would be equivalent to a net release of 553 Tg of CO_2 by 2000 and 4,558 Tg by 2030. Thus, the organic carbon released through land desertification in China could be an important factor affecting changes in concentrations of greenhouse gases worldwide (Feng et al., 2001).

范文 2

At three study sites, representing Mediterranean, semi-arid, and mildly-arid climatic conditions, the effect of shrubs on the spatial patterns of soil moisture was studied. At each site, soil moisture was measured, on hillslopes, at the vicinity of eight shrubs. For each shrub, the measurements have been taken at three microenvironments, i.e. under the shrub (US), at the margins of shrub (MS), and between shrubs (BS). At the microenvironments US and MS, the measurements were taken at three directions: upslope, downslope, and sideslope of the shrubs. At all sampling points, soil samples were taken from three depths: 0~2 cm, 2~5 cm and 5~10 cm. In addition, rock fragments cover percentage near the shrubs was determined. A soil moisture pattern was found, around each shrub, which is composed of a radial gradient and a downslope gradient. The radial gradient is expressed by soil moisture decreasing from the US microenvironment, in all directions, through the MS towards the BS microenvironment. The US microenvironment has a 'spatial advantage' of higher soil moisture content due to (1) relatively higher infiltration rate, (2) capture overland flow from the BS area upslope that shrub, and (3) low evaporation rate because of the shading effect. The downslope gradient is expressed by decreasing soil moisture from the upslope direction of each shrub (MS and US microenvironments) towards the downslope direction of that shrub (MS and US microenvironments, respectively). This gradient is controlled by the relatively high content of rock fragments near the shrubs at their upslope direction. Such rock fragments spatial distribution is attributed to (1) The detachment and transport of rock fragments by sheep and goats trampling and (2) the effect of shrub on the continuity of overland flow and sediment transport. The effect of rock fragments is similar to that of shrubs regarding increasing infiltration and decreasing evaporation rate. The relatively high soil moisture at the upslope direction of each shrub enhances annuals growth producing a positive feedback loop: soil moisture-annuals growth-trampling. This sequence maintains the typical rock fragments spatial organization and contributes to the sustainability of the grazing system. At all the study sites at the US microenvironment, there is a trend of decreasing soil moisture with increasing soil depth. At microenvironments, MS and BS soil moisture increases with soil depth. The results are of great relevance for rehabilitation strategies as they suggest that in order to combat desertification in degraded semi-arid and mildly-arid areas, where the main land use is grazing, both shrubs and rock fragment should be kept at their present spatial distribution (Pariente, 2002).

范文 3

Sandstorms are catastrophic weather phenomena that occur frequently in arid and semi-arid areas with enormous effects on the ecological environment and human health. Particularly around oases, desertification is the major obstacle to sustainable development of oases in arid regions of Northwest China. At present, the problem of how to build an optimal oasis-protection system must be resolved by local farmers and scientists. This paper used data on the sand transportation rate, wind speed, and the width, and water requirement of vegetative barriers to evaluate the shielding effect of an oasis-protection system. The results showed that integrating straw checkerboards, a sand-fixing plant belt, and a farmland shelter belt provided an effective oasis-protection system in the transitional zone between an oasis and shifting dunes. The system can reduce wind speed by 70% and the sand transportation rate by 96% when wind reached the farmland after passing through the oasis-protection system. For the transitional zone between the oasis and a Gobi desert, a desert vegetation preservation belt, a windbreak belt, a mixed belt that combines production forest and forage grass, and a farmland shelter belt also provided significant shelter, and reduced wind speed and the sand transportation rate by 70% and 80%, respectively. For an optimal oasis-protection system, the width and water requirements of the vegetation barriers were also crucial factors because water resources are an important limiting factor for maintaining vegetation barriers in the long-term (Zhao et al., 2008).

范文 4

Reactive oxygen species (ROS) play a key role in the acclimation process of plants to abiotic stress. They primary function as signal transduction molecules that regulate different pathways during plant acclimation to stress, but are also toxic byproducts of stress metabolism. Because each subcellular compartment in plants contains its own set of ROS-producing and ROS-scavenging pathways, the steady-state level of ROS, as well as the redox state of each compartment, is different at any given time giving rise to a distinct signature of ROS levels at the different compartments of the cell. Here we review recent studies on the role of ROS in abiotic stress in plants, and propose that different abiotic stresses, such as drought, heat, salinity and high light, result in different ROS signatures that determine the specificity of the acclimation response and help tailor it to the exact stress the plant encounters. We further address the role of ROS in the acclimation of plants to stress combination as well as the role of ROS in mediating rapid systemic signaling during abiotic stress. We conclude that as long as cells maintain high enough energy reserves to detoxify ROS, ROS is beneficial to plants during abiotic stress enabling them to adjust their metabolism and mount a proper acclimation response (Choudhury et al., 2017).

范文 5

Biodiversity experiments have shown that species loss reduces ecosystem functioning in grassland. To test whether this result can be extrapolated to forests, the main contributors to terrestrial primary productivity, requires large-scale experiments. We manipulated tree species richness by planting more than 150,000 trees in plots with 1 to 16 species. Simulating multiple extinction scenarios, we found that richness strongly increased stand-level productivity. After 8 years, 16-species mixtures had accumulated over twice the amount of carbon found in average monocultures and similar amounts as those of two commercial monocultures. Species richness effects were strongly associated with functional and phylogenetic diversity. A shrub addition treatment reduced tree productivity, but this reduction was smaller at high shrub species richness. Our results encourage multispecies afforestation strategies to restore biodiversity and mitigate climate change (Huang et al., 2018).

范文 6

Despite the wide spread media attention of chain-reaction traffic incidents and property damage caused by windblown dust in the U. S. and elsewhere in the world, very few studies have provided in-depth analysis on this issue. Remote sensing and field observations reveal that wind erosion in the Southwestern U. S. typically occurs in localized source areas, characterized as "hot spots" while most of the landscape is not eroding. In this study, we identified the spatial and temporal distribution patterns of hot spots that may contribute dust blowing onto highways in the Southwestern U. S.. We further classified the hot spots for the potential of blowing dust production based upon field observations and wind erosion modeling. Results of land use and land cover show that shrubland, grassland, and cropland accounted for 42%, 31%, and 21% of the overall study area, respectively. However, of the 620 total hot spots identified, 164 (26%), 141 (22%), and 234 (38%) are located on shrubland, grassland, and cropland, respectively. Barren land represented 0.9% of the land area but 8% of the dust hot spots. While a majority of these hot spots are located close to highways, we focused on 55 of them, which are located <1 km adjacent highways and accessible via non-private roads. Field investigations and laboratory analysis showed that soils at these hot spot sites are dominated by sand and silt particles with threshold shear velocities ranging from $0.17 \sim 0.78$ m/s, largely depending on the land use of the hot spot sites. Dust emission modeling showed that 13 hot spot sites could produce annual emissions N 3.79 kg·m^2, yielding highly hazardous dust emissions to ground transportation with visibility <200 m. Results of location, timing, and magnitude of the dust production at the hot spots are critical information for highway authorities to make informed and timely management decisions when wind events strike (Li et al., 2018).

范文 7

Desertification is a change in soil properties, vegetation or climate, which results in a persistent loss of ecosystem services that are fundamental to sustaining life. Desertification affects large dryland areas around the world and is a major cause of stress in human societies. Here we review recent research on the drivers, feedbacks, and impacts of desertification. A multidisciplinary approach to understanding the drivers and feedbacks of global desertification motivated by our increasing need to improve global food production and to sustainably manage ecosystems in the context of climate change. Classic desertification theories look at this process as a transition between stable states in bistable ecosystem dynamics. Climate change (i. e. aridification) and land use dynamics are the major drivers of an ecosystem shift to a desertified (or degraded) state. This shift is typically sustained by positive feedbacks which stabilize the system in the new state. Desertification feedbacks may involve land degradation processes (e. g. nutrient loss or salinization), changes in rainfall regime resulting from land-atmosphere interactions (e. g. precipitation recycling, dust emissions), or changes in plant community composition (e. g. shrub encroachment, decrease in vegetation cover). We analyze each of these feedback mechanisms and discuss their possible enhancement by interactions

> with socio-economic drivers. Large scale effects of desertification include the emigration of "environmental refugees" displaced from degraded areas, climatic changes, and the alteration of global biogeochemical cycles resulting from the emission and long-range transport of fine mineral dust. Recent research has identified some possible early warning signs of desertification, which can be used as indicators of resilience loss and imminent shift to desert-like conditions. We conclude with a brief discussion on some desertification control strategies implemented in different regions around the world (D'Odorico et al., 2013).

11.4 Introduction

论文(研究报告)的引言为读者提供了论文(报告)的方向,为他们提供了理解后面章节中详细信息所需的视角。引言的撰写可分为5个阶段(步骤)(图11-1)。

在引言的第一阶段,作者应该撰写包括关于研究领域的一般性陈述,为读者提供论文(报告)科学问题产生的背景。在第二阶段,作者应就该科学问题领域内已经研究过的问题的各个方面提供更具体的陈述。在第三阶段,作者应表明对拟调查(研究)的科学问题需要进行更多文献的补充和查阅(以体现该研究的必要性)。在第四阶段,作者应对研究目的(目标)做出非常具体的陈述。第五阶段为可选的陈述,即为即将开展研究的价值。

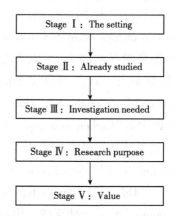

图11-1 The basic stages of writing an introduction of a research report

作者并不总是严格按照以上顺序安排引言写作的各个阶段。有时,作者可以用另一个阶段来打断一个阶段,然后返回到上一个阶段。有时,第二阶段(通常称为"review of literature")与引言的其余部分完全分离。例如,在硕士和博士学位论文中,它通常写成一个单独的章节。第五阶段经常被完全省略。这里给出的论文撰写的常用方案,对于研究论文的初学者(作者)来说是最容易使用的。

(1) Stage Ⅰ

第一阶段的主要目的是为读者提供必要的背景,让他们了解作者研究的特定主题与一般研究领域的关系。首先要对作者所从事的领域做出明显的、普遍接受的陈述。换而言之,在第一阶段,作者建立了一个背景或参考框架,以帮助读者理解该研究如何融入更广泛的研究领域。所以,作者应从一些与一般领域相关的公认的事实开始展开背景介绍。在一般区域内先确定一个子区域,最后指出研究的主题,采取循序渐进的方法让读者更接近研究主题。作者可以通过几句话或几段话来完成这个目标,或者可以通过新旧信息顺序将论文写作思路联系起来。为了引导读者顺利地理解第一阶段的写作思路，作者利用新旧信息将句子联系起来。这可以采用将旧信息（即读者已知的信息）放在句子的开头，并将新信息放在结尾的方法来实现这个过程。

在引言撰写的第一阶段，作者通常应当使用一般和特定的名词短语。由于引言的第一阶段通常始于一般领域的事实陈述，这里也包括作者研究的特定主题。当作者撰写这些类型的一般陈述时，通常使用指代最普遍水平的对象或概念的名词。当句子中包含涉及一整类事物的名词时，应该使用通用名词短语来表达这个意思。

类属名词是指一类具有特定属性的所有名词的统称。例如，"Psammophytic shrubs" "Ecosystem carbon sequestration"等概念。在英语中，一般名词短语有不同的写法。如果这个名词是可数的，可以加上复数标记"s"并省略任何冠词，或者使用无限冠词"a"或"an"将其与单数形式一起使用，使其通用。例如，"Arid and semi-arid ecosystems, covering over 40% of the terrestrial surface, are expected to be highly sensitive to climate and land-use changes"中的"ecosystems"和"changes"就是泛指的例子。又如，"An invasive species can change ecosystem processes"中的"an"加上"species"泛指任何一种入侵物种。当想用的名词不可数时，可以省略任何一个冠词使其成为通用名词。当然，不可数名词不带复数标记"s"。例如，"Carbon exchange between the terrestrial biosphere and the atmosphere is one of the key processes that need to be assessed in the context of the Kyoto Protocol"中的"Carbon exchange"就是一个不可数名词的使用而表示泛指。

单数形式的可数名词有时与定冠词"the"一起使用时具有一般意义。例如，"The hummingbird can be found in all areas of North America"中的"hummingbird"前加"the"表示泛指。"The United States has experienced the integration of the computer into society"中的"computer"也是一种泛指。引言撰写的第一阶段，通常包含使用很大比例的通用名词短语。对于特指名词短语的表达，在稍后的引言撰写中，有必要参考或使用特定的项目和概念，以便将读者从一般领域转移到作者所研究的特定科学问题上来。这里需要使用名词短语，其中名词指的是一类的特定个体成员的总称，而不是整个大类。因此，如果作者指的是指定事物的假定或共享信息，便可以使用定冠词"the"。例如，"In recent years the growth of desert areas has been accelerating in the world"中"the"的使用强调了沙漠面积的增长。

使用冠词"the"的另一种情况是旧信息（之前已经提及的信息）的指向。因此，当提及已经提及的特定事物时，可以使用定冠词"the"，而第一次提及时通常使用不定冠词"a"或"an"。例如，"New Mexico Solar Energy Institute is developing a computerized diagnostic assistant for solar domestic hot water systems. The computer-implemented assistant will be used at naval shore facilities throughout the world."中的第二个"the"是指第二次提及的"a computerized diagnostic assistant"。当作者指向指定信息时，当在随后的短语或子句中明确了具体含义时，可以使用定冠词"the"。例如，"The desert plant species which survived under the severe drought are all characterized by deep roots and high root system ratio"中特别指出了"drought"对植物物种的影响。

当写作时，请问自己一个问题——这个名词是一般意义上的还是指特定意义上的。如果是特指的，请在名词前使用"the"。如果是一般性指向的，再问自己一个后续问题——这个名词是可数的还是不可数的。如果它是可数的，在末尾使用"a"或"an"（单数）或"s"（复数）。如果它是不可数的，则什么也不用(不加冠词或"s"的复数标记)。

当作者表达前期提过的信息时，可以使用单词的重复和派生。例如，"Approximately three years ago, an apparently new and unexplained disorder called acquired immune deficiency syndrome (AIDS) was recognized. Characteristically, AIDS is associated with a progressive depletion of T cells."中"AIDS"就是一种重复。另一个例子是这样的，"It has significant of interest that part of the world's ice which occurs on rivers. Although river ice forms only a fraction of the total quantity of ice in the world"。另一种表达之前提过信息的方式是使用代词和指示词。例如，"Water is one of the most intriguing substances on earth. It has the interesting property that its freezing point is within the range of the earth's surface temperature variation for significant parts of the year"。另一个例子是"Ice forms when water is cooled to 0℃ and continues to lose heat. Generally, this happens when the air temperature falls below 0℃"。

实现提及前期提到信息的另一种方法是使用时间序列的方式。例如，"Curly top virus can be a serious problem in tomatoes. The incidence (of curly top virus) varies from year-to-year"。另一个例子是"With holding or with drawing life-supporting treatment is one of the most important ethical issues for medicine in the late twentieth century. At least six physicians have been accused of murder (to give you one example of the ethical consequences involved in with holding or with drawing treatment) this year alone."是通过(事件的先后顺序)举一个例子，说明拒绝或停止维持生命的治疗所涉及的伦理后果。

总之，当撰写引言的第一阶段时，信息的陈述应该从一般性陈述转向具体陈述。作者应该从某一研究领域普遍接受的事实陈述开始，在研究主题的一般区域中确定一个子区域。按逻辑顺序组织论文的思路。在句子开头使用已发表的信息。在这一阶段的写作语言中，作者应该适当地标记通用名词短语的复数标记"s"、"a"或"an"、"no"、冠词"the"。通过使用重复或派生的单词、代词、指示词或暗示的方式来表达之前提及的信息。

(2) Stage Ⅱ

在第二阶段，作者应回顾感兴趣的研究领域已发表的研究结果(发现)。因此，第二阶段通常被称为"review of literature"(文献综述)。它本质上是对其他研究结果的参考文献或引文的的组合与排列，这些参考文献或引用在论文(报告)末尾中单独列出。它还延续了第一阶段，为读者提供了解作者研究所需的背景信息，并向读者阐明研究所在领域进行此项研究的重要性。第二阶段的内容将作者的研究确立为该领域一系列研究中的一个环节，作者的研究正在发展和扩大该领域的知识。

例如，"In most deserts of the world, transitions between topographic elements are abrupt and watercourses which are dry most of the time tend to dissipate their occasional waters within local basins. Occasional torrential rainfall, characteristic of most desert regions, washes loose debris into watercourses or transports this material, depositing it in and along the shores of ephemeral lakes. These physical processes result in a redistribution of dead plant material (litter), affect the distribution of soil water and create a heterogeneous biotic community. Therefore, before the dynamics of desert ecosystems can be adequately understood, the spatial relationships must be clarified. There have been few studies of litter distribution and/or soil fauna in any of the world deserts (Wallwork, 1976). Wood (1971) surveyed the soil fauna in a number of Australian arid and semiarid ecosystems. Wallwork (1972) made some studies of the micro arthropod fauna in the California Mojave desert. Edney et al., (1974, 1975, 1976) studied abundance and distribution of soil micro-arthropods in the Mojave desert in Nevada. In the Chihuahuan desert, Whitford et al., (1975, 1976, 1977) described the spatial relationships for many groups of organisms, but soil micro-arthropods remain unstudied. The lack of such information represents a gap in our knowledge of desert ecosystems. As part of our continuing program of the designed study reported here is to understand the relationship between litter redistribution and the spatial distribution and composition of the soil micro-arthropod community"。

当引用其他文献时，可以选择突出该作者提供的信息，或突出关注作者本人。前者称之为信息突出的引用，因为信息被赋予了首要的重要性。以下是突出引用信息的例子："In most deserts of the world, transitions between topographic elements are abrupt" "The literature on teaching effectiveness has established few theoretical grounds to guide the selection of meaningful variables"。

另一种类型的信息突出引用使用括号内的数字编号(而不是作者的姓名和日期)。编号是指论文末尾按数字或字母顺序编号的参考文献列表。例如，"The introduction of high strength, high flexibility materials has raised the need for a dynamic approach to floor design[1, 2, etc.]"。信息突出的引文通常用于表示引言第二阶段的开始。在引言第二阶段，指的是你研究的一般领域的研究结果(发现)，它们也可能出现在第一阶段。在信息突出的引文中，当作者需要强调引文的信息时，应该采用现在时态撰写。当作者引用的信息是普遍接受的科学事实时，就使用现在时态。随着"review of literature"(文献综述)的继续，引文中的内容是与作者研究更密切相关的研究内容。在这类的引文中，作者的名字被给予了更多的关注。此时，作者的姓名是句子的主体，后面括号中标注日期或引文编号，然后是引用的信息。这种引文称为作者突出型引用。例如，"Leopold (1921) listed foods, but gave no quantitative data"。作者突出引用的时态较弱。

当引用重点是几位作者的研究领域时，使用现在完成时。这种引文称为弱化作者信息型的引用。在引言第二阶段的后期，可以使用作者突出的引用来报道与自己的研究密切相关的个人研究结果。在这些引文中，论文(报告)的动词使用一般过去时。

现在完成时也用于描述一个领域内研究工作水平的一般陈述。这些陈述往往没有引用。信息突出的引用、弱化作者的引用和一般陈述通常写在第二阶段的开头或第二阶段内新章节开头的过渡点。

在第二阶段的"review of litterature"(文献综述)中，选择的焦点有助于确定动词的时态。同样，在突出作者的引用中，对研究人员研究结果(发现)的态度也会影响第二阶段句子中补语动词的形式。因此，一项特定研究的结果(发现)可以被普遍认为是事实或某项特定研究的结果仅限于某研究，这两种情况下的研究结果并非在所有情况下都是成立的。当你引用的研究的作者可能对他们的研究结果和发现感到犹豫，或者他们可能根本没有完全报道研究结果，而只是提出一些意见或建议的引文时。如果你相信你引用的研究结果(发现)是事实时，在补语动词中使用现在时[即句子中给出研究结果(发现)的部分的动词]。如果你认为这些研究结果(发现)仅限于你引用的具体研究时，在补语动词中使用过去时。如果你引用的研究结果(发现)被原作者视为犹豫的，或者只是意见或建议而不是确定的研究结果(发现)，在报道的动词中使用助动词，并在补充动词中使用语气助词。上述3种情况，论文(报告)的动词都是过去时，而研究结果(发现)部分的动词时态将会因作者的态度而异。

总之，应使用一个确定的逻辑顺序来撰写引言并排列信息。在第二阶段的开始和过渡点使用信息突出和弱化作者而突出引用文献的价值。使用作者突出的文献来报道第二阶段后期的具体研究结果。使用现在时来表示事实陈述，使用现在完成时来弱化作者的文献引用和其他研究的一般陈述。使用过去时来表示作者突出的引用和仅限于单个研究的结果。使用语气助词来表示意见或建议。在补语中使用语气助词来表示初步的(不确定的)研究结果(发现)。

(3) Stage Ⅲ, Ⅳ, and Ⅴ

当你介绍了背景设置并讨论了其他研究人员已发表的工作后，你可以使用引言的最后部分将读者的注意力集中在你将在论文(报告正文)中所要处理的具体科学问题上。这是由另外三个额外的阶段完成的，我们将其指定为Ⅲ、Ⅳ和Ⅴ。阶段Ⅲ表示一个在以前的文献中没有研究过的领域，但从你自己研究的角度来看，这是很重要的。第四阶段正式指出你的研究目的。第五阶段表示你的研究工作可能带来的利益或应用前景。最后，对第Ⅲ、Ⅳ和Ⅴ阶段中包含的信息类型进行有逻辑性的排列组合，使读者从"review of literature"(文献综述)中了解你的研究目的。

第三阶段提供了(研究领域)缺失的信息，因为它向读者发出了已经完成撰写"review of literature"(文献综述)的信号。它通过总结文献指出一个研究的空白来，也就是其他作者没有研究过的一个重要研究领域。通常，第三阶段只需一两句话就能完成。如何指出前人研究留下的研究差距可以在第四阶段回答，因为第四阶段旨在尽可能简明扼要地说明了论文(研究报告)的具体目标。在这个阶段，可以直接陈述研究的目的，因为它满足了第三阶段对你研究领域的额外研究需求的补充。研究目的陈述的方向可能是论文(报告)本身，也可能是指表达研究内容的论文(硕士学位论文、博士学位论文或研究报告)。或者，研究目的陈述的方向可能是研究工作本身，换句话说，研究工作本身，而不是论文(报告)。

在第五阶段，根据这项研究工作可能对该领域的其他研究人员或在实际工作的人具有的一些价值或应用价值来体现你研究工作的合理性。我们可以把这个阶段称为研究价值陈述。第五阶段并不包括在每一个引言中。通常，当你写硕士学位论文、博士学位论文或项目申请书时，你应该在引言中包含第五阶段。当陈述用外部资金开展的项目(结题)报告时，价值陈述通常包含在其中以体现研究的价值。作为期刊论文的撰写中，第五阶段经常被省略。

价值的陈述可以从研究的结果可能带来的实际利益的角度来撰写，也可以强调研究结果在提高特定研究领域的知识水平的理论重要性方面来写价值陈述。

当把引言的最后三个阶段中的内容都撰写完成后，在每种情况下所做的选择将决定撰写这些阶段所需的词汇和语法结构。特殊的信号词通常用于指示阶段 III 的开始。诸如"however"连接词的使用就是用于此目的。连接词后面紧跟着现在或现在完成时的间隙语句，该语句通常包含"few, little, or no"的修饰语。如果隐含的研究问题是"是"或"否"问题，连接词"whether or if"在第四阶段使用，语气助词像"will"或"could"伴随动词。当隐含问题是信息问题时，省略"if（whether)"并使用不定式或名词短语。

第五阶段的价值陈述，作者通常以一种暗示、谨慎或谦逊的态度来书写。在撰写研究的论文(报告)时，不应该对研究工作所产生的无论是实际的还是理论的价值听起来太确定。这很正常，听起来更确定的就应该更加谨慎。这是在第五阶段通过使用情态动词来实现的，主要是"may"。

11.5 Materials and methods

11.5.1 Materials

尽管论文(实验研究报告)的第二个主要部分通常被称为"method"，但有时也称为"materials and methods"。这一组合标题表明研究人员通常将这两个方面的内容放在一起描述，也就是说，他们同时描述了开展实验的每个步骤中使用的任何设备(实验仪器)或其他材料。

如果使用的材料是所在领域研究人员所熟知的，那么传统的做法是只提及即可。然而，如果在实验中使用了特殊设计或非常规材料，通常在论文(报告)中对其进行详细描述。这种情况下，应该按以下顺序安排信息。首先，作者需要提供一个由一两句话组成的整体概述，使读者对材料和目的有一个大致的了解。其次，需要提供材料主要部分的描述，每个主要部分或特征都是按逻辑顺序描述的。最后，需要开展材料的功能描述，以显示第二步中描述的各种功能是如何协同工作的。

实验材料的整体概述如以下范文："The see-saw dryer was developed for the drying of coffee and cocoa beans. It was intended for small-scale drying operations and could be easily operated. It was designed for use in tropical regions. The dryer was operated in two positions along a central axis of rotation running north-south. This see-saw operation permitted the drying material to face the sun more directly during both morning and afternoon."

介绍实验材料的原理如以下范文："The dryer consisted of a rectangular wood frame divided lengthwise into parallel channels of equal width, and crosswise by means of retaining bars. The bottom of the dryer was made of bamboo matting painted black. The cover of the frame was made of a film of transparent polyvinyl chloride (P. V. C.) which provided a screening effect against ultra violet light, thus reducing photodegradation of the drying product. All of the internal parts of the dryer were coated with a flat black paint."最后，所用材料的功能表示为"The drying frame was tilted during operation so that it faced east during the morning and west during the afternoon"。

当提供实验材料的主要特征时，根据实验材料，可以按空间排列的方式进行组织安排，如从上到下、从前到后、从左到右、从中心到外部，也可以其他空间方式来描述特征。这种布置对于描述由各种连接部件组成的设备特别有用。还可以按照功能进行组织安排，如按功能顺序从头至尾描述主要功能。这种安排适用于描述按固定顺序运行的部件。

通常可以将实验材料与实验过程或方法结合起来进行描述。研究中使用的实验材料有时与实验过程分开描述。当使用几种不同的常规实验设备开展常规性实验时，可以使用这种组织安排。例如，"All the aromatic compounds used were commercially available materials without further purification. 2-propanol was distilled from sodium metal. The instrumentation used included an HFT-80 and NT-300 spectrometer, a Hewlett Packard 5980-A mass spectrometer, a Waters Associates HPLC Instrument with a 10-ft column containing 15% Carbowax on Chromosorb W"。然而，更常见的是，材料和方法是以综合的形式展开描述的，通常在每一句话中都提到这两个要素。例如，"Aqueous sodium hydroxide (30 g, 185 mL) was cooled in ice in a 500 mL beaker, stirred magnetically while 5 g of nickel-aluminum alloy added in several small portions, and gradually warmed to 100℃ as required to maintain the hydrogen evolution. The nickel was then allowed to settle and the liquid was decanted. After being washed with 5% fresh sodium hydroxide and distilled water until neutral, the nickel suspension was filtered with a glass funnel and then finally washed with 100 mL of 2-propanol to a glass-stoppered bottle"。

在语言方面，当描述研究中使用的实验样品（本）和材料时，常用过去时。例如，"The boys were between the ages of 7 and 13"。然而，当描述从中选择样本或受试者的一般人群时，通常使用现在时。例如，"All students who apply for admission to the American University of Cairo take The Michigan Test of English Language Proficiency"。如果在使用的实验设备在该领域是标准的或常规的，并且可能是其他大多数研究人员所熟悉的，应该采用现在时态来描述。例如，"The Auditory Test for Language Comprehension permit the assessment of oral language comprehension of English and Spanish"。另外，对于该领域其他研究人员可能不熟悉或特殊设计材料的描述通常采用过去时。对于该研究中使用的以某种特殊方式改造的常见设备的描述，有时也会采用过去时态。例如，"The Michigan Test of English Language Proficiency was Protected from weather by an outer window of 10 mm tedlar"。

主动语态和被动语态动词结构都可用于描述实验材料，选择何种语态部分取决于动词是及物的还是不及物的。

如果动词是及物的，应遵循以下规则来确定语态。例如，"The temperature inside the chamber was increased from 0 to 20℃""Four thermocouples were monitored hourly"。

主动语态通常用于没有人直接负责操纵实验材料的情况，即实验材料"自行"操作的情况。例如，"A 200 hp generator provided power to the piezometers""Control gauges monitored air pressure inside the chamber"。

被动语态可以用来描述涉及非人类主体参与的过程，但必须包含一个短语来指示主体。例如，"Power was supplied by 14 generators with capacities ranging from 90 to 300 kW"。

总之，实验材料中的信息是指将实验材料与实验方法描述相结合的过程。首先简要地提及常规与常见的实验材料，并使用三步顺序描述专门设计的实验材料。在描述主要部分时，可以选择按照空间或功能来排列的方式。在语言方面，描述实验样本时使用过去时。在描述特别设计的实验材料时，使用过去时；当描述常见的实验材料时使用现在时。如果动词是不及物的，并且动作是"自己发生的"，则使用主动语态；如果动词是及物的，并且动作中涉及人为因素，则使用被动语态。实验方法部分描述了作者开展研究时遵循的步骤以及每个步骤中使用的实验材料。它对那些想了解作者研究方法及其如何影响实验结果，或有兴趣重复和拓展该研究的读者很有用。

11.5.2 Methods

方法部分中包含的要素及呈现顺序并不是固定的。通常，方法中包含的信息要素包括"Overview of the experiment""Population/Sample""Location""Restrictions/Limiting conditions""Sampling technique""Materials""Variables"和"Statistical treatment"。例如，"Our 3-year study of changes in the ratio of serum to serum creatinine in Colorado wild bears began in the winter of 1981 and ended in the fall of 1983""The investigation was performed in the Black Mesa-Crystal Creek area in West-central Colorado""The study area has three major vegetation bands: a mountain shrub community at lower elevation (2235 ~ 2330 m), large aspen forests of Engelmann spruce and fir at higher elevations""A total of 76 blood samples were obtained from 27 female and 21 male bears""Bears were captured with Aldrich spring-activated foot and lower leg snares""Snared bears were immobilized with a combination of ketamine hydrochloride""A six-foot pole was used to administer the drug""In winter, the bears were located with a radio signal emitted by the bears collars""The samples were cooled, serum was separated from red blood cells, and urea and creatinine concentrations were determined" "Statistical analysis of changes in blood parameters was done with Scheffe's comparison because seasonal values could not be considered either independent or dependent"。

在描述开展研究所遵循的步骤或实验方法时，内容应该写得非常清楚，这样你所在领域的读者就可以重复开展实验。描述一个过程的最佳方式是循序渐进或按时间顺序的方式进行排列。当对所采用的实验方法和过程进行描述时，通常使用一般过去时。如果方法中引用其他研究者描述的标准或常规方法，而在你的研究中属于特殊方法时，也采用过去时撰写。

在描述项目研究中使用的方法与过程时，可以使用主动语态或被动语态。在方法过程描述中使用主动语态还是被动语态时取决于以下因素。首先，被动语态通常用于描述方法和过程，使信息去个人化。被动结构需要省略主体(通常是"I"或"We")，重点放在方法和过程是如何完成上。除了时态问题外，选择主动语态还是被动语态应该把旧信息放在句子开头，把新信息放在结尾。旧信息在每句话中都用斜体字表示。例如，"The four reactors we tested in the work reported here all contained a platinum catalyst" (主动语态); "Each reactor-catalyst configuration will be described separately (passive)" (被动语态); "The quartz reactors were manufactured by the Wm. A. Sales Company of Wheeling, Illinois (passive)" (被动语态)。在科技论文过程和方法描述写作中，对以下常用的3类句子通过下列方式进行缩短。第一类是对于由两个相同的主语和两个或多个被动动词组成的复合句中，为了缩短这类句子，省略句子后半部分的主语和助词。第二类也是复合句，但在这种情况下有两个不同的主语，每个主语的被动语态都有不同的动词。为了缩短这种句子，省略第二个动词前的"be"助词。第三类句子有一个包含被动动词形式的"which"从句。在这种情况下，可以通过去掉连接"which"和"be"助词来缩短子句。

总之，方法部分用于描述实验开展的过程。该信息应包括有研究者能够重复研究过程中所需的所有信息。在语言方面，可以用过去时来描述实验过程，也可以用被动语态来解除个人化而对过程进行描述，并在句子开头保留旧信息，还可以使用被动语态的简短形式来减少复合句和从句的使用。

11.6 Results

结果部分是论文(实验研究报告)的第三个主要部分,其中介绍了研究结果并对进行了简要评论。一些作者将这一部分称为"Results and discussion",从而对研究结果发表了更有深度的评论。本节只介绍包含简短评论的结果,将独立性的评论置于下一节介绍。在撰写结果部分之前,与拟投稿期刊的《作者指南》核实格式要求。

论文(报告)的结果部分以数字和文字形式介绍了研究结果。图和表格多以数字形式呈现完整的研究结果,而随附的文本有助于读者关注结果的最重要方面并对其进行解释。

结果部分包括3个信息要素:一是用于定位可以找到结果图表的语句;二是陈述最重要的研究结果(发现)的语句;三是对结果发表评论的语句。写作时,可以通过包含上面提到的所有3个要素来替换结果的缩写形式。例如,可以将要素一和要素二组合起来,以呈现最重要结果的语句,并在括号中注明可以找到这些结果的图表。在要素三中,可以添加对结果进行评论和注释的语句。例如,"Caffeine was somewhat more potent than theophylline in preventing leaf-eating. In contrast, caffeine has been reported elsewhere to be ten times weaker than theophylline as an adenosine antagonist"。

这是结果的另一个例子:"A total of 53 samples were examined. Direct microscopic examination of the samples showed 20 different fungal strains, which were isolated by culture and identified to the level of genus and/or species. These findings show that fungi can tolerate adverse environmental changes in the vegetative form. Table 2 shows the results of the physiological tests applied to the isolates. None of the fungi strains was able to grow in culture media with 500~5000 mg/L of anionic surfactant. An inhibitory effect on fungal growth and activity might be expected from the anionic surfactant level found in the ponds"。

有两种可能的方法对结果的评论语句(要素三)进行安排:一种是在提到的每一条重要结果后写一条简短的评论;另一种是在写完所有结果之前进行评论。评论的作用是从结果中概括、解释结果的可能原因,并将结果与其他研究的结果进行比较分析。

在使用三步法撰写结果部分时,应该遵守以下动词时态的用法。在要素一中,使用现在时态来定位图表中的数据。例如,"Results of the T-tests are presented in Table 1" "Table 4 summarizes the test results on precontaminated insulators"。需要注意的是,关于图表位置的内容陈述可以用主动语态或被动语态书写,但在这两种情况下都使用现在时态。

陈述重要的研究结果(要素二)通常使用过去时。例如,"As a group, divorced mothers spent over twice as much time in employment as married mothers (Figure 2)" "The coefficient of correlation was found to be significant at the 0.001 level"。在一些领域,如工程和经济学,作者可能会用现在时表达他们研究的发现。

评论研究结果(要素三)通常使用现在时或语气助词。对论文研究结果与其他研究的结果进行比较并发表评论，通常使用现在时。例如，"This is consistent with earlier findings suggesting that personal characteristics are not related to attrition and teaching"。使用评论对结果给出可能的解释通常使用情态助动词。例如，"These results can/may be explained by considering the voltage distribution on 230 kV insulators during freezing conditions"。通过评论来概括结果的内容通常使用"may"。例如，"Hyperactive children may be generally responsive to amphetamines"。

在第三要素评论中，也可以使用现在时态中试探性动词而不是语气助词从结果中概括结果的内容。例如，"It appears/seems/is likely that hyperactive children are generally responsive to amphetamines" "These results suggest that children who display learning problems are depending on only one cerebral hemisphere"。

要素二提出了不同类型的研究结果。在一些研究中，研究结果涉及各个处理组之间的比较，通常会设置一个或多个处理组与对照组。在这种情况下，要素二语句通常使用比较级或最高级表达式的方式进行撰写。例如，"The professional athletes had faster eye movements than our other subjects"。在其他研究中，研究结果显示了一个变量随时间变动的趋势。当撰写这些类型的结果时，在要素二语句中使用变体表达式或特殊的变体动词。例如，"Prices showed a tendency to increase over the 3-year period" "The percentage tended to decline in the second half of the female students decade"。第三类研究结果显示了一个变量与另一个变量的关系，或变量之间的关系。当你撰写这些类型的结果时，要素二中使用相关或关联动词是很常见的。例如，"Choice of location was correlated with/associated with marital status" "Dry weight of top growth was not highly/significantly/closely related to total nitrogen"。

11.7 Discussion

讨论是论文(报告)的最后一个主要部分。在讨论部分，进一步对研究结果进行全面审视和整体看待。正如在引言中所描述的一样，作者(研究人员)通过讨论部分在相较其研究领域更大背景下审视研究工作。随着讨论部分的逐步推进，作者将读者的注意力从研究的具体结果上转移开，开始更普遍地关注并提出研究对该领域其他工作的重要性。

例如，最初的假设是"The decremental theory of aging led us to infer that older workers in speed jobs would have poor performance, greater absenteeism and more accidents compared with other workers"。研究的结果是"The findings, however, go against the theory. The older workers generally earned more, were absent less, had fewer accidents, and had less turnover than younger workers"。对研究结果的两种解释是"One possible conclusion is that the requirements of the speed jobs in the light manufacturing industry under study do not make physical demands on the older workers to limits of their reserve capacity. The competence and experience of the older workers in these specific jobs may have compensated for their reduced stamina"。

研究结果所产生的限制条件是"This study has taken a step in the direction of defining the relationship between age, experience, and productivity in one particular industry. It is possible of course that other industries with a different complex of speed jobs and skill jobs may produce entirely different results. In addition, it is important to emphasize that methodological problems in the research design limit our interpretations"。进一步开展研究的必要性是"The approach outlined in this study should be replicated in other manufacturing plants, as well as in other occupational areas in light, medium, and heavy industries in order to construct a typology of older worker performance in a variety of jobs"。因此，遵循写作思路的信息排列顺序将作为研究的具体参考。然后，应该添加对研究的主要目的或假设的一个引用。在此之后，应该对最重要的发现进行检查，无论它们是否支持最初的假设，以及它们是否与其他研究人员的研究结果一致。需要安排对这些研究结果（发现）可能产生的解释或推测。研究的另一个局限性限制了研究结果的推广程度，也限制了研究人员对研究结果的立场。最后，需要在讨论的最后列出研究结果的含义（结果的概括）以及对未来研究和实际应用的建议。

在讨论部分，比论文（报告）中的其他任何地方，作者应更明确地表达他们对研究及其研究结果（发现）的看法。作者对研究结果的解释、含义、局限性或应用采取立场性的陈述（要素三）。例如，"One possible explanation is that speed jobs do not tax older workers to their limits"（对研究结果的解释）；"We can no longer assume that it is satisfactory to seek explanations only in economic factors"（对研究结果的含义进行解释）；"We acknowledge that other industries may produce different results"（提出研究所具有的局限性）；"Clearly, this technique has promise as a tool in evaluation of forage"（对研究结果的应用进行描述）。

在讨论部分，可以使用过去时、现在时或情态动词。最常用于指称目的、假设和发现的动词时态是一般过去时。选择取决于对具体发现的解释是仅限于作者的研究（过去），还是指一般情况（现在）。情态动词也可以用来强调这些陈述的推测性质。当将研究结果（发现）与其他研究人员的研究结果（发现）进行比较时，使用现在时作为动词的时态。

当从研究的具体重要性转向整个研究重要性的更广泛、更一般陈述时，使用一般现在时和语气助词或助动词。例如，"It appears that squatter housing markets behave as economically rational entities"中的内容暗示了研究的重要性或含义。另外，"The approach outlined in this study should be replicated in other manufacturing plants"和"We recommend that the approach outlined in this study should be replicated in other manufacturing plants"中的内容显示了该研究提出的建议和市场应用前景。

当用表达式来重述假设时，可以使用加上"that"形式的复合句。例如，"It was anticipated the theory led us to infer""In line with this hypothesis""We assumed""The results seem inconsistent"和"With our hypothesis"。又如，"It was anticipated the theory led us to infer""In line with this hypothesis""The results seem inconsistent"和"With our hypothesis+that+older workers in speed jobs would have poorer performance than younger workers"。当对研究结果做出解释或据此对未来做出一些预测是，可以使用"These findings suggest""imply""lend support to""the assumption lead us to believe/provide evidence""These findings suggest""imply""the assumption lead us to believe/provide evidence+that+frost affects the pan by breaking its massive structure"。

11.8 Acknowledgments

根据所投稿期刊的要求，所有未列入作者行列而需要致谢的人员、资助该研究的项目、基金名称等信息需要在致谢部分列出。

11.9 References

参考文献列表依据所投期刊的格式要求整理。

References

AGHABABAEIAN H, OSTADTAGHIZADEH A, ARDALAN A, et al. , 2021. Global health impacts of dust storms: A systematic review[J]. Environmental Health Insights(15): 1-28.

BAGNOLD R A, 1941. The physics of blown sand and desert dunes[M]. London: Methuen & Co. Ltd.

BOFAH K K, KRAMER C, GERHARDT H J, 1991. Design considerations for buildings in a sandy and dusty environment [J].Journal of Wind Engineering and Industrial Aerodynamics, 38: 2-3.

CHEN G, WANG W, SUN C, et al., 2012. 3D numerical simulation of wind flow behind a new porous fence [J]. Powder Technology, 230: 118-126.

CHEN Y, CAI Q, TANG H, 2003. Dust storm as an environmental problem in north China[J]. Environmental Management, 32: 413-417.

CHOUDHURY F K, RIVERO R M, BLUMWALD E, et al., 2017. Reactive oxygen species, abiotic stress and stress combination [J]. The Plant Journal, 90 (5): 856-867.

D' ODORICO P, BHATTACHAN A, DAVIS K F, et al., 2013. Global desertification: drivers and feedbacks [J]. Advances in Water Resources, 51: 326-344.

DEL PRETE S, GIULIO I, MARIO P, et al., 2010. Origin and distribution of different types of sinkholes in the plain areas of Southern Italy [J].Geodinamica Acta, 23(1-3): 113-127.

DONG Z, LUO W, QIAN G, et al., 2007. A wind tunnel simulation of the mean velocity fields behind upright porous fences [J]. Agricultural and Forest Meteorology, 146(1-2): 82-93.

DONG Z, QIAN G, LUO W, et al., 2006. Threshold velocity for wind erosion: the effects of porous fences [J]. Environmental Geology, 51(3): 471-475.

FENG Q, CHENG G, MIKAMI M, 2001. The carbon cycle of sandy lands in China and its global significance [J]. Climatic Change, 48(4): 535-549.

FORSTER E J, HEALEY J R, DYMOND C, et al., 2021. Commercial afforestation can deliver effective climate change mitigation under multiple decarbonisation pathways [J]. Nature Communications, doi: 10.1038/S41467-021-24084-X.

HILLEL D, ROSENZWEIG C, 2002. Advances in agronomy [M] // SPARKS D L, ed. Desertification in relation to climate variability and change. Beijing: Academic Press.

HUANG Y, CHEN Y, CASTRO-IZAGUIRRE N, et al., 2018. Impacts of species richness on productivity in a large-scale subtropical forest experiment [J]. Science, 362 (6410): 80-83.

JIANG Z C, LIAN Y Q, QIN X Q, 2014. Rocky desertification in Southwest China: Impacts, causes, and restoration[J]. Earth-Science Reviews, 132(3): 1-12.

Kassam A, Derpsch R, Friedrich T, 2014. Global achievements in soil and water conservation: The case of conservation agriculture [J]. International Soil and Water Conservation Research, 2 (1): 5-13.

LI J, KANDAKJI T, LEE J A, et al., 2018. Blowing dust and highway safety in the southwestern United States: Characteristics of dust emission "hotspots" and management implications [J]. Science of the Total Environment, 621: 1023-1032.

MIDDLETON N J, 2017. Desert dust hazards: A global review[J]. Aeolian Research, 24: 53-63.

MOHAMMED A E, STIGTER C J, ADAM H S, 1996. On shelterbelt design for combating sand invasion[J]. Agriculture, Ecosystems & Environment, 57(2-3): 81-90.

NICKLING W G, NEUMAN M K, 2009. Aeolian sediment transport [M]. Berlin: Springer.

PARIENTE S, 2002. Spatial patterns of soil moisture as affected by shrubs, in different climatic conditions [J]. Environmental Monitoring and Assessment (3): 237-251.

QU J, ZU R, ZHANG K, et al., 2007. Field observations on the protective effect of semi-buried checkerboard sand barriers [J]. Geomorphology, 88 (1-2): 193-200.

RAINE J K, STEVENSON D C, 1977. Wind protection by model fences in a simulated atmospheric boundary layer[J]. Journal of Wind Engineering and Industrial Aerodynamics, 2(2): 17-39.

TARNITA C E, BONACHELA JA, SHEFFER E, et al., 2017. A theoretical foundation for multi-scale regular vegetation patterns[J]. Nature, 541(7637): 398-401.

VIGIAK O, STERK G, WARREN A, et al., 2003. Spatial modeling of wind speed around windbreaks[J]. Catena, 52(3-4): 273-288.

WANG T, 2011. Deserts and aeolian desertification in china [M]. Beijing: Science Press.

WASSON R J, HYDE R, 1984. Factors determining desert dune type (reply) [J]. Nature, 309(5963): 92.

WERNER B T, 1990. A steady-state model of wind-blown sand transport [J]. The Journal of Geology, 98(1): 1-17.

YAO Z, XIAO J, MA X, 2021. The impact of large-scale afforestation on ecological environment in the Gobi region[J]. Scientific Repports, 11(1): 14383. https://doi.org/10.1038/s41598-021-93948-5.

YUAN D X, 1997. Rock desertification in the subtropical karst of south China [J]. Zeitschrift für Geomorphologie, 108: 81-90.

ZHANG Z X, REES H W, DONG Y, et al., 2014. Assessment of effects of two runoff control engineering practices on soil water and plant growth for afforestation in a semi-arid area after 10 years [J]. Ecological Engineering, 64: 430-442.

ZHAO W, HU G, ZHANG Z, et al., 2008. Shielding effect of oasis-protection systems composed of various forms of wind break on sand fixation in an arid region: A case study in the Hexi Corridor, northwest China [J]. Ecological Engineering, 33(2): 119-125.

Appendix

A

ablique crestal ridge 斜脊沙垄
above-surface weathering 面上风化[作用]
above tide 海拔
abrasion 磨蚀[作用]
abrasion plane 磨蚀面
absolute drought 绝对干旱
absolute humidity 绝对湿度
absolute maximum 极端最高
absolute minimum 极端最低
absolute moisture 绝对含水量
abtuse bimodal wind regime 钝双峰风况
accelerated erosion 加速侵蚀
accelerated weathering 加速风化[作用]
accidented relief 崎岖地形,起伏地形
accommodation 调节[作用],适应[作用]
accretion 外加堆积,停滞堆积
accumulated temperature 积温
accumulation 堆积[作用]
action of sheet erosion 片蚀作用
active aeolian deposition 活动风成沉积
active dune 活动沙丘,流动沙丘
active porosity 有效孔隙度
active sand sea 活动沙海,流动沙海
active sand dune field 活动沙丘原;流动沙漠
active wind, erosion wind, sandwind 活动风,侵蚀风,沙风
actual sand-flow rate 实际沙流速率,实际输沙流速率
acute bimodal wind regime 锐双峰风况
adierstein 风成盆地
adsorbability 吸附性,吸附能力
adsorption 吸附[作用]
advancing dune 前移沙丘
aeolian accumulation 风成堆积
aeolian activity 风成活动
aeolian basin 风成盆地
aeolian clay deposit 风成黏土沉积物
aeolian deflation 风蚀
aeolian deposit 风成沉积物,
aeolian dune 风成沙丘
aeolian dust 风成尘土
aeolian effectiveness 风成效率
aeolian erosion 风蚀
aeolian form 风成形态,风成地貌
aeolian geomorphology 风成地貌学
aeolian landform 风成地貌
aeolian landform of sand 风沙地貌,风成沙地貌
aeolian layer 风成层
aeolian material 风成物质
aeolian movement 风成移动,风沙移动
aeolian ripple 风成沙纹
aeolian sand 风成沙
aeolian sand-size parameter 风成沙粒级参数
aeolian sand stream 风成沙流,风沙流
aeolian sediment, aeolian sedimentary deposit 风积物
aeolian sedimentary particle 风积颗粒
aeolian sedimentary process 风成沉积过程
aeolian sedimentary structure 风成沉积构造
aeolian soil 风成土,风积土,风沙土
aeolian system 风成系统
aeolian transport 风成般运
aeolianist 风成学家,风成论者

aeolium 风积物，风积物层
aerodynamic ripple 空气动力波纹，空气动力沙纹
afforestation 人工造林
afforestation of sands 沙地造林，固沙造林
agba 山口，垭口
aggressive water 侵入水，侵蚀性水
aghurd dune 金字塔沙丘，山状沙丘
agricultural afforestation 农田防护林营造，护田林营造
agricultural amelioration 农业土壤改良
air-dry 风干
air dry weight 风干重
air drying 风干，凉干
air gap, air-gap 风口，干峡口
air permeability 通气性
air subsidence 空气下沉
airborne particulate 空中悬浮微粒
air-flow, airstream, air stream 气流
air-mass 气团
alab dune 线形沙丘
alab longitudinale 纵向线形沙丘
alluvial basin 冲积盆地
alluvial deposit 冲积物
alluvial erosion 冲积侵蚀
alluvial fan 冲积扇
alluvial flat 河漫滩
alluvial meadow 冲积草甸
alluvial sand 冲积沙
alluvial soil 冲积土
alluvial terrace 冲积阶地
alluviation 冲积[作用]
alluvium（复数 alluvia）冲积层，冲积物
alpine desert 高山荒漠
alpine meadow 高山草甸
altitude 高度，海拔
alveolar barkhan sands 窝状新月形沙丘地（综合沙垄）
alveolar dune 窝状沙丘，格状沙丘

alveolar weathering 凹窝风化，蜂窝状风化
amelioration 改良，土壤改良
ammocolous 沙生的
ammophilous 喜沙的，沙生的
amount of sand drift 输沙量
amount of sand passing 过沙量
amphitheatre 圆形凹地；风蚀坑
ancient aeolian environment 古风成环境
angle of internal friction 内摩擦角
angle of repose 休止角
angle of sliding friction 滑动摩擦角
angle of slope, slope angle 坡度角
angle of static friction 静态摩擦角
annual resultant drift direction 年合成输沙方向
anthropic erosion 人为侵蚀
anthropogenic desertification 人为荒漠化
anthropogenic monitoring 人为活动监测
anthropogenic vegetation 人工植被
anticyclone 反气旋
anticyclone circulation 反气旋式环流
anticyclone subsidence 反气旋下沉
anticyclonicity 反气旋作用
anti-erodibility 抗蚀性
antierosion measure 防蚀措施
antimosoon 反季风
anti-trades, counter-trades 反信风
apex 冲积扇顶；顶点
aqabah, aqubaht 山口，垭口
aquatic ecosystem 水域生态系统，水生生态系统
aquatic habitats 水生生境
aquatic soil 水成土
aqueduct 输水管，引水槽
aqueo-residual sand 水蚀残沙
aqueous deposit 水成沉积
aquifer, aquafer 含水层，蓄水层
aquifuge 隔水层，隔水岩体，不透水岩石系
arc 新月形体，新月形段
arch 拱顶，穹窿

area of sand deposit 积沙区
areal degradation 区域剥蚀
arenaceous 沙质的
arenose 粗沙质
arenosol 砂土
argillaceous 泥质的
argillaceous desert 黏土荒漠，土漠
arheism 无流区(无径流和水系)
arid and semi-arid lands (ASAL) 干旱和半干旱土地，干旱和半干旱地区
arid desert 干旱荒漠
arid erosion 干旱侵蚀
arid landforms 干旱地形，干旱地貌
arid part, arid regions, arid territory 干旱区域
arid steppe 干草原
arid morphostucture 干旱形态结构，干旱地貌结构
aridisol 旱成土
aridity 干旱性，干燥性
aridity coefficient 干燥系数
aridity index 干燥指数
aridization 干旱化
aridization of land 土地干旱化
arrested dune 稳定沙丘，固定沙丘
arm of dune 丘臂，指抛物线沙丘和星状沙丘的向外延伸的臂
arm of sand rose 沙玫瑰臂
armoured surface 砾石面，砾石防护面
arrested dune 稳定沙丘，固定沙丘
artificial dune 人造沙丘
artificial rain 人工降雨，人造雨
artificial vegetation 人工植被
arvideserta 流动沙漠群落
aspect of slope, slope aspect 坡向
asymmetric ridge 不对称沙垄
asymmetrical barchan 不对称新月形沙丘
asymmetrical dune form 不对称沙丘形态
asymmetrical wind regime 不对称风况

atmospheric circulation 大气环流
atmospheric turbulence 大气湍流
attrition 磨蚀
authigenic sand 自生沙(由海水直接沉积产生的沙)
available soil moisture 有效土壤水
available precipitation 有效降水
available water 有效水，可给态水
available water holding capacity 有效持水量
avalanche deposit 崩塌沉积物，滑塌沉积物，塌沙堆积
avalanche face 崩塌面，滑塌面
avalanche slope 滑塌坡
avalanching 塌沙[作用]，崩塌[作用]
avalanching slipface 崩塌滑落面
average grain diameter 平均粒径
average moisture content 平均含水量
AWS = automatic weather station 自动气象站
azonal phenomenon 非地带现象
azonic 非地带性的
azotification 固氮[作用]

B

bab 通道，走廊，山口，垭口
backing of wind 风向逆转
backwearing 后退侵蚀[作用]
Badain Jaran Sha-mo 巴丹吉林沙漠
Badlands, bad lands 崎岖荒凉地，劣地，侵蚀地
baghara(复数 baghair)沙丘，流动沙丘
balance of hydrological inputs and outputs 水文收支平衡
bank 岸；滩；沙洲
bank erosion 堤岸侵蚀
barchan, barcan, barchane, barkan, barkhan 新月形沙丘
barchan chain 新月形沙丘链
barchan swarm 新月形沙丘群
barchanoid draa 新月形类大沙丘

barchanoid element 新月形段
barchanoid ridge 新月形沙垄
barchanoid ridge coalescing 新月形沙垄连接
barchanoid section 新月形段
barchanoid-type dune 新月形类型沙丘，新月形类沙丘
barchanoid windward slope 新月形沙丘段迎风坡
brachan-to-lineal model 新月形沙丘向线形沙丘的演变模式
bare dune 裸露沙丘
bare karst 裸露喀斯特
bare transverse dune 裸露横向沙丘
barren land 不毛之地
barren sands 裸沙地
barrial 干盐湖
basal sapping 基部掏蚀
base 山麓，山脚，丘脚，底面
base level of erosion 侵蚀基准面
base line 基线
base of slope 坡脚
basin irrigation 淹灌
bay dune 海湾沙丘
beach erosion 海滩侵蚀
beach sand 海滩沙，湖滩沙
beach-side dune ridge 滩边沙丘垄
bed load 推移质，底沙
bed rougness 河床粗糙度；床面粗糙度
bedload channel 推移质河道，含泥沙河道
beds of precipitation 沉积层
belt of weathering 风化带
berg wind 东风，山风
betterment of land 土地改良
bidirectional wind 双向风
bimodal aeolian sand 双峰风成沙
bimodal wind 双峰风
bimodal-sized sand 双峰粒级沙
bioaccumulate 生物累积
biocoenose, biocommunity 生物群落

bioerosion 生物侵蚀
biogenic deposit 生物沉积
biogenous weathering 生物风化[作用]
biological barrier 生物屏障；生物沙障；活沙障
biological dune stabilization 生物沙丘固定；生物固沙
biological erosion 生物侵蚀
biological fixation 生物固定
biological means 生物措施，生物固沙措施
biological stabilization method 生物固沙方法，生物固土方法
biological weathering 生物风化[作用]
black blizzard, black dusters 黑尘暴
black buran, black storm 黑风暴
black earth 黑土
blair 平原
blanket of aeolian sediment 风积物层
blanket sand 冲积沙，冲积覆盖沙层
blocked drainage 堵塞水系
blocked channel 淤塞河道
blocking action 阻塞作用
blowhole 风蚀穴，风蚀坑
blowing dust 扬尘，吹尘
blown dust 飞尘，扬尘，浮尘
blown land 风蚀地
blown sand 流沙，风沙流，扬沙
blowout dune, blow-out dune 风蚀坑沙丘
blow-out hollow 风蚀凹地
blowout niche 风蚀穴，风蚀坑
body of water 水体，水域
booming sand 鸣山，响沙
boreas 北风
boulder clay 泥砾
boulder-field 巨砾原，戈壁滩
boulder pavement 砾漠
bouncing 反跳
boundary ridge 边界沙垄
bowl 碗状洼地；盆状落水洞

bra, bras 沙丘链，沙丘垄，纵向沙丘链，纵向沙丘垄
brash 风化碎石
breakdown 塌崩
breed 小沙丘
brown desert soil 棕漠土
brown earth 棕壤
bulk density 容重
bulk transport 容积搬运
bush-mound 灌丛沙堆
by-oasis sand 近绿洲沙地，绿洲缘沙地
by-stream deposit 近河沉积[物]

C

cachment area 集水区
cachment basin 集水盆地
calcipit 汇水盆地，汇水洼地
caldrom 小洼地，锅状盆地
calm belt, calm zone 无风带
calves 小沙丘
calving 崩解[作用]
canal 运河；渠道
canal aqueduct 输水渡槽，输水渠道
canyon, canon 峡谷
canyon wind 下降风，下吹风
capillarity 毛管作用
capillary adsorbed water 毛管吸附水
capillary capacity 毛管持水量
capillary condensation water 毛管凝结水
capillary force 毛管力
capillary front 毛管上限
capillary humidity 毛管水
capillary interstice 毛管孔隙
capillary migration 毛管[水]运行
capillary pore 毛管孔隙
capillary porosity 毛管孔隙度
capillary rise 毛管上升
capillary tension 毛管张力

capillary water-holding capacity 毛管持水量
captation 沙阱，陷阱积沙
captation dune 不定形沙丘
cardinal wind 盛行风
carrying capacity 承载力；输沙量；载畜量，载牧量
carrying capacity of the environment 环境负荷量，环境承载力
catagenesis 后退演化，退化
catch basin 贮水池；汇水盆地
catchment area 汇水区，集水区
catchment basin, drainage area 流域盆地
cavernous weathering 洞穴风化
caving 成洞[作用]；河岸冲刷；坍塌
caving zone 坍塌区，下沉区
cellular soil 多孔状土壤，蜂窝状土壤
chain 山脉，山岭，沙丘链
chain of connected dune 连接沙丘链(新月形沙丘链)
chain of star 星状沙丘链
channel basin 河谷盆地
channel deposit 河道沉积[物]
check dam 拦水坝
checkerboard fence 格状沙障，格状栅栏
chemical corrosion 化学溶蚀[作用]
chemical denudation 化学剥蚀[作用]
chemical deposit 化学沉积物
chemical desertification 化学荒漠化
chemical disintegration 化学崩解[作用]
chemical erosion 化学侵蚀
chemical sand stabilization 化学固沙
chemical soil stabilizer 化学土壤固定剂，化学固定剂
chemical solution 化学溶蚀
chemical weathering 化学风化
chemical sand stabilization 化学固沙
chevron dune V形沙丘
chomoeremion 砾漠

cistern 蓄水池
clay aggregate 黏粒团聚体
clay content 黏土含量
clay desert 黏土荒漠，土漠
clay dune 黏土沙丘
clay fraction, clay size, clay-size fraction 黏粒粒级
clay matrix 黏土基质
clay mineral 黏土矿物
clay particle 黏粒
clay plain 黏土平原
clay sand 黏质沙土
clay silty loam 黏质粉沙壤土
clay dune sediment 黏土沙丘沉积物
clay enriched horizon 黏土富集层
clayey beach-dune ridge 黏质海岸沙丘垄
clayey desert 黏土荒漠
clayey lunette 黏土沙丘
clayey sand 黏土质沙
claying horizon 黏化层
clay-rich environment 富黏土环境
clay-rich sediment 富黏土沉积物
clay-rich subsurface horizon 黏土富集亚表层
clean sand 纯沙
climatic degeneration, climatic deterioration 气候恶化
climatic fluctuation 气候变动
climatic oscillation 气候波动
climatic shift 气候变动，气候变迁
climatic variation 气候变迁
climatron 人工气候室
climax dune type 巨型沙丘类型
climbing dune 爬行沙丘
clough, claugh 沟，狭谷
cluster dome 簇状穹丘
cluster of compound parabolic dune 综合抛物线沙丘簇
cluster of star duns 星状沙丘簇

coacervation 团聚[作用]
course alluvium 粗粒冲积物
coarse particle 粗颗粒
coarse clay 粗黏粒
coarse crumb 粗屑粒
coarse dust 粗尘
coarse fraction 粗粒粒级
coarse grain 粗(颗)粒
coarse grained soil 粗粒质土壤
coarse gravel 粗砾
coarse gravel deposit 粗砾沉积物
coarse sand 粗砂
coarse sand sheet 粗砂沙片
coarse sandy loam 粗砂壤土
coarse sandy soil 粗砂质土壤
coarse silt 粗粉沙
coarse soil 粗质土
coarse textured soil 粗粒质土壤
coarse weathered mantle 粗粒风化层
coastal sand belt 海岸沙带
coastal sand dune 海岸沙丘
cobble 粗砾
cobble soil 粗砾质土
cockscome ridge 风蚀残垄，风蚀土脊
coefficient of friction 摩擦系数
coefficient of permeability 透水系数
cohesionless particles 非黏结性颗粒
cohesive sediment 黏结性沉积物
cold arid lands 寒带干旱土地，寒带干旱地区
cold climate dune 冷气候沙丘
cold current 寒流
cold desert 寒漠
cold hardiness 耐寒力
cold resistance 耐寒性
collapse 塌陷，崩塌
colluvial deposit 崩积物
colluvial deposition 崩积沉积
colluviation 崩积作用

complex barchan 复合新月形沙丘
complex crescentic dune 复合新月形沙丘
compound crescentic dune ridge 综合新月形沙丘垄
complex dune 复合沙丘
complex linear dune 复合线形沙丘
complex longitudinal dune 复合纵向沙丘
complex parabolic dune 复合抛物线沙丘
complex sheet 复合沙片
complex star dune 复合星状沙丘
complex streak 复合沙带
complex transverse dune 复合横向沙丘
composite dune 复合沙丘
compound barchan 综合新月形沙丘
compound barchanoid dune 综合新月形沙丘
compound barchanoid ridge 综合新月形沙垄
compound dome shaped dune 综合穹状沙丘
compound feathered linear dune 综合羽状线形沙丘
compound linear dune 综合线形沙丘
compound mound 综合沙丘，综合沙堆
compound parabolic dune 综合抛物线沙丘
compound ridge 综合沙垄
compound sheet 综合沙片
compound streak 综合沙带
compound U-shaped dune 综合U形沙丘
comprehensive amelioration 综合土壤改良
concentration of flow 集流
conduit 输水管，输水渠，引水渠
cone of depression 沉陷锥；下降漏斗（指地下水）
cone of depression 下降漏斗
confined flow 承压水流
confined ground water 承压地下水
confined water 承压水
confining bed (layer, stratum) 不透水层，隔水层
confluence 汇流点，汇流处，汇流河口
conical dune 圆锥形沙丘（金字塔沙丘）
conical hill 锥形丘，锥形小山

conical pile of sand 圆锥形沙堆
conservation of soil and water 水土保持
consolidate 固结
consolidation 固结［作用］
constant wind 稳定风
constructional plain 堆积平原
consumptive use 总蒸发量，总耗水量
contamination 污染
contour 等高线；等深线
contour line 等高线
convection 对流
convective storm 对流暴雨
conventional grain-size curve 通用粒度曲线
coppice dune 灌丛沙堆
coppice mound, coppice dune 灌丛沙堆，灌丛沙丘
corrosion 溶蚀［作用］，腐蚀［作用］
corrosion resistance 耐腐性
corrosiveness 溶蚀性；腐蚀性
corrosivity 溶蚀性；腐蚀性
coupole 圆顶丘
cover degree 覆盖度
coverage 盖度，覆盖度
crack 裂隙
creep 蠕移
creep fraction 蠕移粒级
creep load 蠕移质
creep transport 蠕移搬运
creeping grain 蠕移颗粒
crescent dune 新月形沙丘
crescent lake 新月形湖，弓形湖
crescent ridge 新月形沙垄
crescent segment 新月形段，新月形体
crescent spring 月牙泉
crescentic dune 新月形沙丘
crescentic dune ridge 新月形沙丘
crescentic element 新月形部位；新月形段
crescentic form 新月形形态

crescentic ripple 新月形波纹
crescentic-shaped hollow 新月形状凹地
crescentic-shaped mound 新月形状沙堆
crest 丘脊(沙丘的最高部位);山脊,峰;顶部
crestal area 丘脊部位,丘脊区
crestal configuration 丘脊轮廓
crest/brink area 丘脊/丘檐部位,丘脊/丘檐区
crest-to-crest distance 丘间距离,丘脊间距
crevice water 裂隙水
critical angle 临界角,体止角
critical drag 临界阻力,临界拖力
critical ecological zones 脆弱生态带,临界生态带
critical humidity 临界湿度
critical point 临界点
critical shear stress, critical shearing stress 临界切应力
critical surface-roughness 临界地面粗糙度常数
critical temperature 临界温度
crust 结壳,结皮,地壳
cultivated vegetation 人工植被,栽培植被
cumulative frequency distribution curve 累积频率分布曲线
cumulative grain size frequency curves 累积粒度频率曲线
cumulative temperature 积温

D

daf 干河床,干谷
dahna 洼地,低地
dam 坝
dammed lake 堰塞湖
damp sand 湿沙
dasht 荒漠,盐漠,平原
dead valley 死谷,干谷
dead water 死水,停滞水
debris cone 冲积锥,冲积扇
debris flow deposit 泥石流沉积物
decay 风化,衰变

declining sheetflow 减速片流
deep ground water 深层地下水
deepening 加深,向下侵蚀
deflation 风蚀
deflation armor 风蚀残积层,戈壁
deflation ripple 风蚀沙纹
deflational activity 风蚀活动
deflational reg 风蚀性砾漠
deforestation 滥伐森林
deformation of river bed 河床变迁
degraded forest 退化林,衰退林
degraded pasture 退化草原
degree of abrasion 磨蚀度
degree of aeration 透气度
degree of compaction 坚实度
degree of porosity 孔隙度
degree of roundness 磨圆度
degree of salinity 盐渍化程度
degree of saturation 饱和度
degree of slope 坡度
degree of sorting 分选程度
degree of surface roughness 地面粗糙度
degree of weathering 风化度
dehydration, dehydrolysis 脱水[作用]
delta deposit 三角洲沉积[物],三角洲冲积[物]
denudation 剥蚀[作用]
depe 丘陵
depletion 耗竭,退化
deposition 沉积[作用]
deposition behind dam 坝后淤积
deposition by flood 洪水淤积
deposition cycle 堆积循环
deposition interface 沉积界面
depositional erosion 沉积侵蚀
depositional landform 沉积地形
depositional terrace 堆积阶地
depositional topography 堆积地形
depressed area 低洼地

depth of penetration 渗透深度
depth of precipitation 降水深度
depth of weathering 风化深度
dereliction 冲积作用，加积作用
derivation 偏向，偏角；导流，引水
descending water 渗透水，下渗水
descent of water 水渗透
desert 荒漠
desert aeolian sands 荒漠风成沙漠
desert armour 砾漠，戈壁
desert belt 荒漠带
desert clay plain 荒漠黏土平原，土漠
desert creep 荒漠蔓延
desert crust 砾漠
desert denudation 荒漠剥蚀[作用]
desert deposit 荒漠沉积物
desert devil 荒漠尘旋风，荒漠尘卷风
desert drainage system 荒漠水系，荒漠排水系统
desert dry valley 荒漠干谷
desert dune 荒漠沙丘
desert encroachment 荒漠入侵、荒漠前移
desert erosion 荒漠侵蚀
desert erosion feature 荒漠侵蚀景观，荒漠侵蚀地形
desert landform 荒漠地貌
desert mosaic 荒漠砾石覆盖层，砾漠
desert pavement 砾漠
desert plateau 荒漠高原
desert scrub 荒漠灌丛
desert sand rose 荒漠沙玫瑰
desert steppe 荒漠草原
desert storm 荒漠风暴
desert zones of moderate belt 温带荒漠带
deserta 荒漠群落
desertification 荒漠化
desertification control 防治荒漠化，荒漠化防治
desertification hazard 荒漠化危险
desertification trend 荒漠化趋势

desertified land 荒漠化土地
desert-like region 荒漠状地区
desiccation 变干
destocking 停止放牧，禁牧
destruction 侵蚀[作用]，破坏[作用]
destruction of soil 土壤破坏[作用]，土壤侵蚀[作用]
destructional benches 侵蚀阶地
destructional landscape 侵蚀景观
deteriorated range 退化草原
deterioration 退化，衰退；变质
devastation 毁坏，破坏，荒废
devegetation 植被破坏
development of land 土地开发
devil 尘卷风，尘旋风
devolution 崩坍；退化
diffused runoff 扩散径流
disforest 毁林[开荒]
dilapidation 崩落
diluvial deposit 洪积物
diluvial soil 洪积土壤
diluvium 洪积层
direct transport 直接搬运
direction of dune longation 沙丘延长方向
direction of sand transport 输沙方向，风沙移动方向
disafforestation 毁林[开荒]
disaggregated/dispersed clay particle 分散黏粒
disaggregated/dispersed silt particle 分散粉粒
disaggregation 分散[作用]
discharge 流量，流入量
discharge rate of sediment 输沙率
disintegration 崩解[作用]，衰变，裂变，蜕变
dispersed substance 分散物质
dispersion 分散[作用]
dispersion medium 分散介质
dissolution 溶解[作用]
dissolving 溶解，溶蚀

distal area[冲积扇]冲积扇末端区
ditch irrigation 沟灌
divarication 河流分支
diversion 改道(河流)
diversion type fence 导风型栅栏，导风型高立式沙障
divide line 分水线
dividing crest, dividing ridge 分水岭 dome 穹丘，圆丘
dome dune, dome shaped dune 穹状沙丘
dome summit 穹丘顶
dominant wind 盛行风
doming 穹隆作用
doras 壤土
double barchan dune 双新月形沙丘
down wind 下风面，向下风
down-cutting, vertical erosion 下切，垂直侵蚀
downcutting 下切，向下侵蚀
down-cutting erosion 下切侵蚀
downslope 下行坡(沙丘)，背风坡
downwind 下风，顺风
DP. = drift potential 输沙势
draa-sized barchan 巨型链状新月形沙丘
drag 拖曳；阻力
drag velocity 摩阻流速，摩阻速度
dragging arm 拖曳臂
drainage 排水；水系
drainage catchment 流域盆地
drainage channel 排泄水道
draw well 取水井
dried out water body 干涸水体，干涸水域
drift sand 流沙，飞沙
drifting dust 浮尘，飘尘
drip irrigation 滴灌
drop in water surface 水面降落
drought hardiness 耐旱性，抗旱性
drought hazard, dry damage 旱灾，旱害
drought resistance 抗旱性，耐旱性

drought tolerance 耐旱性
dry air mass 干气团
dry matter 干物质
dry meadow 干草甸
dry monsoon 干季风
dry period, dry phase, dry spell 干期，干旱期
dry playa, clay plain 干盆地，黏土平原
dry steppe 干草原
dry streambed 干河床
dry wash 干河床，石砾质河床
drying and wetting cycle 干湿交替
drying and wetting effect 干湿效应
drying effect 干燥效应
drying power 干燥能力
drying sand 干沙
drying up 干涸
dryland, dry land 旱地
dryland salting 旱地盐渍化
dumi 矮灌丛
dune 沙丘
dune activity 沙丘活动性，沙丘流动性
dune arm 沙丘臂(抛物线沙丘)
dune assemblage 沙丘集合体
dune base 沙丘基部，沙丘底面
dune chain 沙丘链
dune cluster 沙丘簇
dune crest 丘脊，丘顶
dune de conjunction 连接沙丘，接合沙丘(纵向沙垄)
dune form, dune-form 沙丘形态
dune mass 沙丘群
dune migration 沙丘移动
dune mound 沙丘
dune network 格状沙丘
dune of longitudinal type 纵向类型沙丘
dune of transversal type 横向类型沙丘
dune plantation 沙丘造林，沙丘人工林
dune pyramidal 金字塔沙丘

dune quadrillage 格状沙丘
dune reticule 格状沙丘
dune ridge 沙丘垄
dune scrub 沙丘灌丛
dune spacing 沙丘间距
dune stabilization 沙丘固定
dune succession 沙丘演化
dune trend 沙丘走向
dune triplets 三连沙丘(新月形沙丘链)
dune tumuli 沙堆
dune wandering 流动沙丘
dunebinder 固定沙丘植物
dune-extradune depositional system 沙丘-丘外沉积系统
dune-free desert pavement 无沙丘砾漠
dune interdune depositional system 沙丘—丘间沉积系统
dust blowing, dust drift 浮尘
dust content 含尘量
dust deposition 尘埃沉积,尘土沉积
dust devil, dust whirl 旋风尘柱,尘卷,尘卷风
dust devil wind 尘卷风
dust fall 落尘,降尘[量]
dust fog 尘雾
dust haze 尘霾
dust mantle 尘覆盖物
dust particle 尘粒
dust source 尘源[区]
dust storm 尘暴
dust turbidity 尘土浑浊度
dust whirl 尘卷,尘卷风
dustiness 尘埃污染[度]
dust-size particle 尘粒级颗粒
dynamic friction 动力摩擦
dynamic threshold 动力起动值
dysgeogenous 不易风化(成碎屑)的

E

E by S = east by south 东偏南
E by N = east by north 东偏北
earth attraction 地球引力
earth flow 泥流
east longitude 东经
east northeast (ENE) 东北偏东
east side 东坡
easterlies 东风带
eastward 向东
ecocide 生态灭绝
ecoclimatology 生态气候学
ecocrisis 生态危机
ecocycle, ecocycling 生态循环
eco-hydrological model 生态水文模型
eco-hydrology 生态水文学
ecological balance 生态平衡
ecological catastrophe 生态灾难
ecological damage 生态破坏,生态损害
ecological desert 生态荒漠
ecological disaster zones 生态灾难带
ecological destroy 生态破坏
ecological disturbance 生态失调
ecological equivalence 生态均衡
ecological equilibrium 生态平衡
ecological fitness 生态适应性
ecological gradient 生态梯度
ecological impact 生态影响,生态冲击
ecological imperative 生态责任
ecological optimization 生态优化
ecological rehabilitation 生态重建,生态恢复
ecological resiliency 生态弹性,生态恢复力
ecological security 生态安全
ecological stability 生态稳定性
ecological succession 生态演替
economic forest 经济林
ecopedology 生态土壤学
ecosphere 生态圈,生态界
ecosystem stability 生态系统稳定性
ecotone 生态脆弱带;群落交错区,群落过渡区

ecoulment 重力滑坡
eddy 旋涡，涡流
eddy current 涡流
eddy flow 涡流，旋流
eddy motion 涡流运动
edge 边，边缘；棱
effect of soil dry 土壤干燥效应
effective abstractions 有效吸水量
effective aperture 有效孔隙
effective moisture 有效水分
effective permeability 有效渗透性
effective pore space 有效孔隙
effective porosity 有效孔隙度
effective precipitation 有效降水量
effective sand-moving wind 有效起沙风
effective wind 有效风
efficiency of irrigation 灌溉效率
efficiency of water application 水分利用率，用水率
efflation 悬浮侵蚀
effluence 流出[物]；侧流
effluent river/stream 潜水补给河
effluent seepage 坡面渗流
efflux velocity 渗出速度，流出速度
effusion 溢流；渗出液
EI Nino current 厄尔尼诺流
elevation 高程，高度；高地
elluvium, eluvium, eluvial horizon 淋溶层，残积层
elongate blowout dune 长风蚀坑沙丘
eluviated clay 迁移黏粒
eluviated grain 迁移颗粒
eluviation 淋溶[作用]；粗化[作用]
embryonal dune 雏形沙丘
encroachment 遇阻堆积；入侵，侵占；前没，漫入
encrustation 板结[作用]，结皮[作用]
endogenic force 内营力

endorheic basin 内流盆地
endorheic drainage 内源水系，内陆河流域
endorheism 内陆流域
energy balance 能量平衡
enrichment 浓缩；富集
entrenched channel 嵌入水道，深切水道
entrenched feeder channel 嵌入补给水道
entrenched stream 嵌入河
environment appraisal 环境评价的
environment contamination 环境污染
environment degradation 环境退化
environment deterioration 环境退化，环境恶化
environment disruption 环境破坏，环境失调
environment evaluation 环境评价
environment hazard 环境危害，环境公害
environment pollution 环境污染
environment protection 环境保护
environment quality 环境质量
environment stability 环境稳定性
environment transition 环境变迁
environment variation 环境变迁，环境变化
environmental appraisal 环境评价，环境鉴定
environmental assessment 环境评价
environmental capacity 环境容量
environmental conservation 环境保护
environmental crisis 环境危机
environmental degradation 环境退化
environmental forecasting 环境预测，环境预报
environmental hazard 环境公害，生态灾害
environmental improvement 环境改善
environmental optimization 环境优化
environmental reconstruction 环境重建
eolation 风蚀[作用]
eolian activity 风成活动
eolian cumulative relief 风成堆积地形
eolian damage 风沙灾害，风成灾害
eolian deposit 风积物
eolian deposition 风成沉积

eolian dune 风成沙丘
eolian environment 风成环境
eolian erosion 风蚀
eolian landform 风成地貌
eolian landforms of sands 风沙地貌，风成沙地貌
eolian sand 风成沙
eolian sand body 风成沙体
eolian sand feature 风沙地貌，风成沙地貌
eolian sand ripple 风成沙纹
eolian soil 风成土，风沙土
eolian suspension 风成悬浮[作用]
eolian transport 风成搬运
eolian-type sandstone 风成类型砂岩
eolomotion 风成沙蠕动
eoposition 遇阻沉积
ephemeral stream 季节性河流，时令河
ephemeral water body 季节性水域，季节性水体
eremology 荒漠学
eroded field 侵蚀地
eroded soil 侵蚀土壤
erodibility 侵蚀度，可侵蚀性，易蚀性
erodible channel 侵蚀河槽
erodible fraction 易蚀颗粒
erodible soil 易蚀土壤
eroding force 侵蚀力
erosion 侵蚀
erosion action 侵蚀作用
erosion agent 侵蚀营力
erosion basin 侵蚀盆地
erosion basis 侵蚀基面
erosion by splash 溅击侵蚀
erosion control 侵蚀防治，侵蚀治理
erosion feature 侵蚀景观，侵蚀地貌
erosion hazard 侵蚀危害，侵蚀灾害
erosion intensity 侵蚀强度
erosion landform 侵蚀地貌
erosion mark 侵蚀痕
erosion pedestal 侵蚀残柱

erosion pits 侵蚀坑
erosion ratio 侵蚀率
erosion resistance 抗蚀性
erosion stream bed 侵蚀河床
erosion/erosional terrace 侵蚀阶地
erosion-aggradation 侵蚀—加积作用
erosional basin 侵蚀盆地
erosional detachment 剥蚀作用
erosional landform 侵蚀地貌
erosional portion 风蚀部位
erosive power 侵蚀能力
erosive velocity 侵蚀速率
erosiveness 侵蚀性
erosivity 侵蚀度，侵蚀性
established dune 固定沙丘
etch 侵蚀，腐蚀
etch pit 蚀坑
etched groove 蚀沟
etching 刻蚀[作用]，腐蚀[作用]
evaporation 蒸发
evaporation from land 地面蒸发(蒸散)
evaporation loss 蒸发耗损
evaporation-rainfall ratio 蒸发—降水比率
evaporative capacity 蒸发量
evaporative discharge 蒸发量，蒸发消耗
evaporative power 蒸发率
evaporative precipitation 蒸发沉淀[作用]
evaporativity 蒸发率
excavation 掘蚀，淘蚀，掏蚀
excavation wind 掏蚀风
exhumed monadnock 裸露残丘
exoreic drainage 外流水系
exotic stream 外源河
exposure 坡向，朝向；露头
extinction 消失(河流)；绝灭(生物)
extra arid desert 超干旱荒漠
extra-catchment area 外集水区，外汇水区
extreme deserts 极端干旱荒漠

extreme water regime 极端水分状况
extremely arid region 极端干旱地区
extremely over-grazing 极度放牧
extremum 极值(最大，最小)
exudation 渗出[作用]

F

facilitate afforestion 封山育林
facing direction 朝向，面向
fall-in 陷落
falling dune 瀑布沙丘
falling water table 下降水位
fallout winds 沉降风
fall wind 下降风
fan 冲积扇
fan apex 冲积扇顶端
fan bay 冲积扇顶
fan head, fanhead 扇顶区
fan mesa 冲积扇残体，冲积扇残丘
fan terrace 冲积扇阶地
fanhead deposit 扇顶沉积物
fan-topped pediments 冲积扇覆盖，山前侵蚀面
fast-moving dune 快速移动沙丘
fast-travelling dune 速移沙丘
fatigue of soil 土壤耗竭
feathered dune 羽状沙丘
feathered linear megadune 巨型羽状线形沙丘
feathered ridge 羽状沙垄
feebly arid 轻度干旱的
fetch 风区，风区长度，吹程
fetch length 风区长度，风程距离
field coefficient of permeability 田间渗水系数
field moisture capacity 田间持水量
field permeability coefficient 田间渗水系数
fill 淤积；充填(洞穴)；洞底淤积泥沙
film water 薄膜水
filtration 渗滤作用
fine crumb 细屑粒

fine dust 细尘(粒径>0.002 mm)
fine fraction 细粒级
fine grain/particle 细颗粒
fine grained sand 细粒沙，细沙
fine granular 细团粒
fine gravel 细砾(粒径 1~2 mm)
fine mass/material 细粒物质
fine pore 小孔隙
fine sand 细沙
fine sandstone 细粒砂岩
fine sandy clayey soil 细沙质黏质土
fine sandy loam 细沙质壤土
fine sandy soil 细沙质土
fine sediment 细粒沉积物
fine silt 细粉沙
fine texture 细质地
fine-grained aeolian material 细岩化风成物质
fine-grained sand 细粒沙
fine grained sediment 细粒沉积物
first order channel 一级水道
first order stream 一级河流
first terrace 一级阶地
fish scale dune 鱼鳞状沙丘
fissure water 裂隙地下水
fixation/stabilization of shifting sand 流沙固定
fixed dune 固定沙丘
fixed ground water, fixed moisture 束缚水
fixed sand desert 固定沙漠
fixing sanddrift 固沙
flash flood, flashflood 暴洪，山洪暴发
flashy stream 湍流河流
floating dust 飘尘，浮尘
flood control 防洪，洪水调节
flood deposition 洪水沉积[作用]
flood erosion 洪水侵蚀
flood flow 洪水径流
flood irrigation 漫灌
flood land 河滩地

flood level 洪水位
flood period 洪水期，汛期
flood protection 防洪
flood regulation 洪水调节
flood retention 拦洪
flood area 受淹区
flow detachment 流水剥蚀，坡面径流侵蚀
flow direction 流向
flow rate 流速
flow separation 气流分离
flow strength 气流强度；径流强度
flow velocity 流速
flowage 流出，泛滥
flow-diversion model 气流转向模式
flowing artesian well 自流井
flowing water 流水
flow-oblique ripple 斜向沙纹
flow-paralled ripple 纵向沙纹
flow-transverse ripple 横向沙纹
fluctuation 脉动，波动
fluctuation of water-table 水位波动，水位升降
fluid 流体
fluid threshold 流体起动值
fluid threshold shear velocity 流体起动切应速度
fluk 纵向沙丘，线形沙丘
flurosion 河流侵蚀[作用]
flusch 松散沉积，松碎沉积
fluvial cycle of denudation 流水剥蚀循环
fluvial cycle of erosion 流水侵蚀循环
fluvial denudation/erosion 流水侵蚀
fluvial deposit 流水沉积[物]
fluvial filtation 流水淤积作用
fluvial landform 流水地貌
fluvial plain 河成平原，冲积平原
fluvial sediment 流水沉积物，冲积物
fluvial system 水系，河系
fluvial transport 河流搬运，流水搬运
fluviation 河成[作用]

fluviative action 河流作用
fluviative deposit 河流沉积[物]
fluviogenic soil 冲积土
flux 流量，通量
forest for conservation of soil and water 水土保持林
forest for conservation of water 水源保持林
forest for wind break and sand stabilization 防风固沙林
forest plantation 植树造林
forest strip 林带，带状林
forest vegetation 森林植被
forest-steppe belt/zone 森林草原带
formative wind 形成风
fractionation 颗粒分选作用
fragmental deposit 碎屑沉积[物]
fragmental soil 砾质土，粗骨土
free dune 自由沙丘
free water 自由水，活水
free water surface, free-water surface 自由水面，活水面
freeze/thaw action 冻融活动
freezing 冻结
freezing and thawing 冻融作用
frequency curve 频率曲线
frequency diagram 频率图
fresh breeze 清劲风(五级风)
fresh gale 强风
freshet 洪水
freshwater, fresh water 淡水
fresh-water deposit 淡水沉积[物]
freshwater lake deposit 淡水湖沉积物
friable soil 松散土壤
friction 摩擦
friction coefficient 摩擦系数
friction velocity 摩擦速度，摩擦风速
frictional force 摩擦力
furrow irrigation 沟灌
fused dune 连接沙丘(新月形沙丘链)

G

gaining stream 地下水补给河流
gale 大风(八级风)
galette dune 饼状沙丘(穹状沙丘)
garmada 抛物线沙丘
garmada dune 抛物线沙丘
gault 重黏土
gein 土壤有机质
gelivation 冰冻[作用]
general base lever or erosion 总侵蚀基准面
geological erosion 地质侵蚀
geomorphological agent 地貌营力
ghurud 线形沙丘
giant barchan 巨型新月形沙丘
giant composite dune 巨型混合沙丘
giant crescent dune pattern 巨型新月形沙丘型式
giant crescentic massif 巨型新月形沙山
giant dome 巨型穹丘
giant parabola 巨型抛物线沙丘
giant ripple, mreye, zibar, giant undulation 巨型沙纹
giant sand massif 巨型沙山
giant undulation 大波状沙垄, 横向波状沙垄
goz soil 砂质土壤
grade 坡度; 粒级
gradient 坡降, 梯度
gradient wind 梯度风
grain diameter 粒径
grading factor 分选系数
grading of sand 沙粒组配, 沙粒递变
grain entrainment 颗粒起动
grain flowage 颗粒流动
grain movement 颗粒移动
grain packing 颗粒填集[作用]
grain shape 颗粒形状
grain size 粒度, 粒径
grain-analyses 粒度分析, 粒级分析

grainfall 颗粒降落, 落尘
grainfall deposit 颗粒降落沉积[物]
grain-size class 粒级
grain-size curve 粒度曲线, 粒级曲线
grain-to-grain collision 粒间碰撞
grait 沙; 卵石
granular structure 团粒结构
granularity 粒度
granulation 团粒化[作用]
granule 团粒(土壤); 细砾(质地)
granule armoured dune 细砾覆盖层沙丘
gravel 砾, 砾石
gravel desert 砾漠, 戈壁
gravel fan 砾质冲积扇
gravel ground/soil 砾质土
gravel mulch 砾漠
gravel pit 砾石坑
gravelly clay 砾质黏土
gravelly coarse sand soil 砾质粗砂土
gravelly fine sand soil 砾质细砂土
gravelly loam 砾质壤土
gravelly plain 砾质平原
gravelly sand soil 砾质砂土
gravelly soil 砾质土
graviplanation 重力夷平[作用]
gravitational/gravity acceleration 重力加速度
gravitational flow 重力流
gravitational slumping 重力滑动
gravitational/gravity water 重力水
gravity irrigation 自流灌溉
gravity slumping 动力滑塌
gravity-controlled slope 重力侵蚀坡
green belt, greenbelt 绿色带
grit 碎石, 砾石, 粗砂岩
grit wave 粗砂沙纹
gritty soil 砂砾质土
gross erosion 总侵蚀量
grouan 粗砂; 砾石

ground depression 地面下沉
ground inversion 地面逆温
ground level weathering 地面风化[作用]
ground water, under-ground water, phreatic water 地下水
ground water discharge 地下水露头，地下水排泄
ground water erosion, groundwater erosion 地下水侵蚀
ground water level 地下水位
ground water outflow 地下水流出(量)
ground water recharge 地下水补给
ground water regression 地下水位下降
ground-level trimming 地面侵蚀
ground-surface subsidence 地面下沉，地面陷落
groundwater basin 地下水盆地
groundwater extraction 抽取地下水
ground-water level 潜水面
groundwater recession 地下水降落
groundwater reservoir 地下水水库
ground-water solution 地下水溶蚀
groundwater storage 地下水贮量
ground-water table 地下水位，潜水位
groundwater withdrawal 抽取地下水
groundwater zone 地下水层，潜水层
gully erosion 沟蚀，冲沟侵蚀
gullying 冲沟作用
gumbo 坚硬黏土
Gurbantunggut Sha-mo, Gurbantonggut Desert 古尔班通古特沙漠
gust 阵风

H

habl 线形沙丘
habitat 生境
hairpin dune U 形沙丘
half-desert 半荒漠
half-shrub 半灌木
haloeremion 盐漠

halomorphic soil 盐碱土
halomorphism 盐成作用，积盐作用
hard water 硬水
headstream 源头，发源地
headward erosion 溯源侵蚀
heat balance 热量平衡
heaving sand 流沙
heavy clay 重黏土
heavy grain 重颗粒
heavy grazing 过牧，过度放牧
heavy rain 大雨(雨量 25.4 mm 以上)
heavy rainstorm 大暴风雨
heavy sand loam 重沙质壤土
high energy effective wind 高能有效风
high latitude 高纬度
high latitude deserts 高纬度荒漠
high latitudes 高纬度地区
high travelling grain 高速移动颗粒
high wind 大风
high wind energy 高风能
high-altitude desert 高山荒漠
high-altitude suspension 高空悬移
high-energy environment 高能环境
high-energy grain 高能颗粒
high-energy sitting 高能沉淀
high-energy wind environment 高能风环境
high-speed wind 高速风
high-water 高潮，高水位，洪水
homogenous sand-trapping system 均质挡沙系统，均质阻沙系统
honeycomb dune 蜂窝状沙丘
honeycomb weathering 蜂窝状风化作用
horizontal transport rate 水平输沙率
horizontal visibility 水平能见度
horn, wing 丘角(新月形沙丘)
horn-to-horn distance 丘角间距(新月形沙丘)
horn-to-horn width 丘角宽度
hudrologic cycle 水分循环

hydration 水化作用
hydration property 水化性质
hydration weathering 水化风化[作用]
hydraulic discharge 涌水量(地下水)
hydraulic force 水力
hydraulic gradient 水力梯度
hydraulic jump 水跃
hydraulic property 水性
hydraulics 水力学
hydrochemistry 水化学
hydrodynamics 流体力学
hydrograph 水文过程线
hydrological cycle 水文循环
hydrolysis 水解[作用]
hydrolytic action 水解作用
hydromelioration 水利土壤改良
hydromorphic soil 水成土
hydromorphism 水成作用
hydro-regime 水分状况
hydroscopic dust 吸湿尘
hydroscopic water/moisture 吸湿水
hydroscopicity 吸湿性
hygroscopic nature 吸湿特性
hygroscopic water 吸着水
hyper aridity 极端干旱(性)
hyper-arid deserts 超干旱荒漠
hyper-arid regions 极端干旱区
hyper-aridity 超干旱性，极端干旱性

I

ice desert 冰漠
illuviation 淀积[作用]
imbricated dune 叠置沙丘
immobile dune 不流动沙丘
immobility 非流动性，不流动性
immobilization 固定作用
impact speed 冲击速度
impact threshold 冲击起动值

impact threshold shear velocity 冲击起动切应速度
impact threshold velocity 冲击起动速度
impeded drainage 排水不良
imperious layer/stratum, impermeable barrier (seam/stratum) 不透水层
impermeable material 不透水物质
impermeability 不透水性
impervious obstacle 不透风障碍物
imperviousness 不透水性，不透气性
imporosity 无孔隙性
impounding sand fence 阻沙栅栏
impoundment 人工湖，蓄水池
improvement of soil 土壤改良
inactive dune 不活动沙丘
incidence dune 斜向沙丘
incoherent slump 松散滑塌
incohesive sand 非黏结沙
index of aridity 干燥指数
index of the directional variability 方向变率指数，风向变率指数
individual dune 个体沙丘
infiltrability 可渗入性，可渗水性
infiltrating water, infiltration water 入渗水
infiltration 入渗[作用]，渗水[作用]
infiltration capacity 入渗量，渗水量
infiltration envelope 渗透层
infiltration rate 入渗速率，渗水速率
infiltration velocity 渗入速度，渗透速度，渗水速度
inflow 流入，注入，水流量
influent 渗流
influent action 渗水作用
influent stream, losing stream 补给地下水的河流
influx rate 渗入速度
inhomogeneity 异质性
inhomogeneous soil 非均质土壤
initial soil 原始土壤
initial surface 原始面

inland basin, interior basin 内陆盆地
inland desert 内陆荒漠
inland drainage 内陆水系
interdune erosion 丘间风蚀
interior desert 内陆荒漠
interlayer water 层间水
intermediate belt 中间带，过渡带
intermediate wind energy 中风能
intermediate-energy wind environment 中能风环境
interstitial flow 渗流
interstitial space/void 粒间孔隙
interstitial water 间隙水
intracapillary space 毛管间孔隙
irrigated area 灌区
irrigated cropland/land 灌溉地
irrigation by flooding 漫灌，大水漫灌
irrigation by gravity 自流灌溉
irrigation cannel 灌溉渠
irrigation ditch 灌渠，灌沟
irrigation water 灌溉水
isarithm 等值线
isobar 等压线
isobaths 等[水]深线
isohume 等湿度线
isohyet 等雨量线
isohypse 等高线
isoombre 等蒸发线
isotherm 等值线
isothermal line 等温线
isothyme 等蒸发量线

J

jump 跳跃；跃移；骤变
jump length 跃移长度
junction 汇流点
junction terrace 汇流阶地

K

kahid 金字塔沙丘

kalut 风蚀垄岗，风蚀长丘
kandi, kani 河流，河道
karst 喀斯特，岩溶
karst erosion 岩溶侵蚀，喀斯特侵蚀
karstification 岩溶作用，喀斯特作用
karstland 喀斯特地区，岩溶地区
kewal 冲积土
klhurd, guern 金字塔形沙丘，塔形沙丘
kinematic friction 动力摩擦
kneaded gravel 泥流搬运砾
knob dune 小沙堆，疙瘩沙丘
knock 小丘，缓丘
knoll 圆丘，小山包
kona 背风面
Korgin Sand 科尔沁沙地

L

lacustrine silt 湖积粉沙
lacustrine transport 湖泊搬运
lag concentrate 粗化物质
lag deposit 粗化沉积物，残留沉积物，风蚀残积物
lag gravel 残留砾石细物质
lag material 残余物质
land aridization 土地干旱化
land classification 土地分类
land control measure 土地管理措施
land degradation 土地退化
land depletion 土地枯竭
land valuation 土地评价
land reclamation 土地改良
land recultivation 土地复耕
land use assessment 土地利用评价
land use capability 土地利用率
land use planning 土地利用规划
land utilization 土地利用
landfrom 地形，地貌
landslide=landslip 滑坡，山崩，塌方，塌坡

landslip 山崩，崩塌，塌方，滑坡
large body of aeolian sand 大风成沙体
large grained 巨粒的，大粒的
lateral corrasion 侧面磨蚀
lateral cutting/erosion 侧蚀
lateral irrigation 侧灌
lateral percolation 侧渗
lateral sedimentation 侧向沉积[作用]
lateral throughflow 侧向渗[透]流
layer erosion 层状侵蚀
leaching 淋溶[作用]
leaching substance 淋溶物质
lee drift 背风坡积沙
lee face, lee-face, lee side, leeside, leeward, leeward side 背风面
leeward slope 背风坡
level of saturation 潜水面，饱和水面，地下水位
level plane 水准面
lichen crust 地衣结皮
lichen weathering 地衣风化[作用]
light clay 轻黏土
light grain, light particle 轻颗粒
light sandy loam 轻沙质壤土
light soil 轻质土
line blow 强风
linear chain 线形沙丘链
linear dune 线形沙丘
linear dune crest 线形沙丘脊
linear erosion 线形侵蚀
linear flow 线形径流
linear loessial hill 线形黄土丘
linear megadune 巨型线形沙丘
littoral deposit 海岸沉积物，湖滩沉积物
littoral dune 海岸沙丘
littoral sand 海岸沙
loaded stream 泥沙河
load-discharge ratio 输沙率
loading 搬运

loam 壤土
loamification 壤质化
loamy fine sand 壤质细砂土
loamy fine soil 壤质细土
loamy sand 壤质砂土
local runoff 地方径流
loess deposit 黄土沉积物
loess desert 黄土荒漠
loess deposition 黄土沉积
longitudinal dune, linear dune, seif 纵向沙丘
loose particle 松散颗粒
low energy effective wind 低能有效风
low energy environment 低能环境
low flow year 枯水年，少水年
low latitude desert 低纬度荒漠
low water level 低水位
lowland sand desert 低地沙漠
low-level transport 低层搬运，低层输送
low-relief desert 低地荒漠
low-velocity flow 低速流
luviation 淋溶作用，淋淀作用
luvic xerosol 淋溶干旱土

M

maceration 浸渍[作用]
macro-dune 巨型沙丘
macropore/macrovoid 大孔隙
macroporosity 大孔隙度
macrorelief 大地形，大起伏
main dune 主沙丘
main ridge 主沙垄
main/major slipface 主滑落面
main river 主河，干流
major stabibized sand seas 固定大沙海
man-made dune 人造沙丘
man-made erosion 人为侵蚀
man-made forest 人工林
marine deposit 海洋沉积[物]

marine sand 海相沙
marine sediment 海洋沉积[物], 海相沉积[物]
mastodon 巨型新月形沙丘
mature dune 壮年沙丘
mature parabolic dune 壮年抛物线沙丘
maximum capillary capacity 最大毛管持水量
maximum capillary rise 最大毛管上升[高度]
maximum instantaneous wend speed 最大瞬时风速
maximum molecular moisture holding capacity 最大分子持水量
meadow 草甸, 低湿草地
meadow bog 草甸沼泽
meal 海滨沙丘带
mealy sand 粉沙
mean annual precipitation 年平均降水量
mean annual runoff 年平均径流[量]
mean depth 平均深度
mean grain diameter 平均粒径
mean grain size 平均粒径, 平均粒度
mean impact velocity 平均撞击速度
mean jump length 平均跃移长度
mean particle 平均粒级, 平均粒度
mean size 平均粒度
mechanical defence 机械防护, 沙障防护
mechanical dune stabilization 机械固沙
mechanical erosion 机械侵蚀
mechanical practice 工程措施
mechanical sand protection fence 机械防沙栅栏
mechanical weathering 机械风化[作用]
median diameter 中值直径
median particle diameter 中值粒径
medium grain 中(颗)粒
medium grain size 中粒径, 中粒级
medium sand 中沙
medium silt 中粉粒
medium size grain 中粒级颗粒
medium-size sands 中沙
megabarchan 巨型新月形沙丘

mega-dune 大沙丘, 巨型沙丘
mega-nebkha 大灌丛沙堆
mega-ripple 巨型沙纹
mega-yardang 大雅丹
melioration 土壤改良
meniscus dune 新月形沙丘
meso-barchan 中新月形沙丘
meso-dune 中沙丘
meso-memory dune 中龄沙丘
microbe(s), microorganism 微生物
microbial process 微生物过程
microbiological corrosion 微生物腐蚀
microbiological crust 微生物结皮
micropore space 微孔隙
microrelief 微起伏, 小地形
microscopic aeolian feature 微观风成特征
micro topography 小地形
migrating dunc 流动沙丘
migrating dune chain 流动沙丘链
migrating/mobilized sand 流沙
migrating sand body 流动沙体
migration 迁徙, 迁移, 位移
migration distance 移动距离, 迁徙距离
migration rate of dune 沙丘移动速率
migratory/mobile dune 流动沙丘
mildly arid region 轻度干旱地区
minimum field capacity 最小田间持水量
minimum water-holding capacity 最小持水量
mini-ripple 小沙纹
minor dune 幼龄沙丘
mixed sands 混合沙
mobile sand desert 流动沙漠
mobility 流动性, 活动性; 迁移率
mobility of element in soil 土壤元素迁移性
mobilization 活化[作用]
moderate rain 中雨(雨量为 12.7~25.4 mm)
moderate salinity 中盐度
moderate energy wind environment 中能风环境

moist air 湿空气
moist air mass 湿气团
moist meadow 湿草地
moist playa 湿盆地
moist season 雨季
moist steppe 湿草原
moisture 水分；水汽；湿润
moisture availability 水分可获得性，水分有效性
moisture balance 水分平衡
moisture capacity 持水量
moisture circulation 水分循环
moisture constant 水分常数
moisture content 含水量
moisture deficiency 水分亏缺
moisture migration 水分移动
moisture tension 水分张力
moisture-holding capacity 持水量
moisture-temperature index 温湿指数
momentum 动量
monadnock 残山，残丘
monitoring 监测
monitory facilities 监测设施
mon-linear negative feature 非线形沉降地形
mon-linear positive feature 非线形上升地形
monsoon, monsoon wind 季风
monsoon air 季风气流
monsoonal circulation 季风环流
mound of sand 沙堆，沙包
mountain desert 山地荒漠，山岳荒漠
mountain mass 山体
mountain pass 山口，垭口
mountain ranges 山脉
mountain ridge 山岭
mountain slope 山坡
mountain system 山系
moving dune 流动沙丘
moving sand 流沙
mudslide 泥石流，泥流

mud rain 泥雨
mudflow deposit 泥流沉积[物]
multi-direction wind 多向风
multi-directional wind regime 多向风况
multi-modal aeolian sand 多峰风成沙
multi-modal wind regime 多峰风况
multiple row belt 多行林带
multi-wind 多向风
musical sand 鸣沙

N

N by E 北偏东
N by W 北偏西
nanophanerophyte 灌木
nanorelief 微地形，微起伏
narrow bimodal wind regime 窄双峰风况
narrow defile 峡谷
narrow shelter belt 窄防护林带
narrow unimodal wind 窄单峰风
natural vegetation 天然植被
natural-anthropogenic balance 自然-人为平衡
natural-anthropogenic regime 自然人为状况
nature preserve 自然保护区
NE by E 东北偏东
NE by N 东北偏北
near desert 半荒漠
near gale 疾风(七级风)
near-ground velocity 近地面风速
near-ground wind 近地面风
nearshore current 近岸流
near surface activity 近地面移动，近地面移动
net sand-moving power 净输沙率
net sand transport potential 净输沙势
net trend of transport drift 净输沙走向
nitrogen fixation 固氮[作用]
nitrogen gathering plant 固氮植物
non-aeolian process 非风成过程
non-aeolian sand 非风成沙

non-capillary porosity 非毛管孔隙度
non-cohesive sand 非黏结沙
non-reversibility 不可逆性
nonstructure soil 无结构土壤
north/northern slope，north-facing slope 北坡
north-east（NE）东北
north-east by east 东北偏东
north-east by north 东北偏北
northeast trades 东北信风
northeaster 东北大风
norther 强北风
northern hemisphere 北半球
northern latitude 北纬
north west（NW）西北
nunja 湿地
nurse crop 覆盖作物
nutrient substance 营养物质
nutrition status 营养状况
nutritive material 营养物质
NW by N 西北偏北
NW by W 西北偏西
nyika 半荒漠；荒原

O

oasefication，oasisofication 绿洲化
oasis 绿洲
oasis sands 绿洲沙地
oblique dune 斜沙丘
oblique flow 斜向气流
oblique linear dune 斜线形沙丘
oblique longitudinal dune 斜纵向沙丘
oblique ridge dune 斜垄沙丘(综合沙垄)
oblique windrift dune 斜风积沙丘(纵向沙垄)
ocean circulation 海洋环流
oghroud 金字塔沙丘，星状沙丘
open ditch 明沟，明渠
open grain 松颗粒
open sand 裸沙，松沙

open-cut drain 明渠，明沟
opposing wind 反向风
opposition dune 反向沙丘
organic matter 有机质
organic matter degradation 有机质降解[作用]
original dune 原始沙丘
original salinization 原生盐渍化
original sediment 原始沉积物
orographic condition 地形条件
orographic precipitation 地形降水，山岳雨
orographic rain 地形雨
osmosis 渗透[作用]
osmotasis 渗透性
osmotic force 渗透力
outflow 流出，流出物
outlet 排水口，出口
outward glow 外向流
over grazing 过度放牧
overcultivation 过度耕作，粗放耕作
over-exploitation 过度开发
overgrazing 过牧，放牧过重
overhead irrigation 人工降雨，喷灌
over-irrigation 大水漫灌，过量灌溉
overland flow 地表漫流
overland runoff 地表径流
overlying dry sand 上覆干沙
overwash 洪积土壤

P

pacticle concentration process 颗粒粗化过程
paddock 围栏分区放牧
paddock grazing 分区轮牧
palaedune，palaeo-dune 古沙丘
palaeo-aeolian deposit 古风成沉积物
palaeo channel 古水道，古河道
parabolic blowout 抛物线风蚀坑沙丘
parabolic dune 抛物线沙丘
partial drought 小旱

particle density 颗粒密度
particle entrainment 颗粒起动，颗粒剥蚀
particle fraction 粒级
particle fractionclassification 粒级分类
particle migration 颗粒移动
particle size, particle-size 粒级，粒径，粒度
particle size analysis 粒级分析，粒度分析
particle size curve 粒度分布曲线
particle trajectory 颗粒轨迹
particulate loading 悬浮微粒[含量]
pass 通道，山口，垭口
pasture rotation 轮牧
pastureland 牧场，草地
patch of sand 沙斑
peak gust 最大阵风
pebbly desert pavement 砾漠
pebbly soil 砾质土
pedocompaction 土壤压缩作用
pelitic texture 黏土质地
pilitization 泥化作用
pellicular moisture/water 薄膜水
penetration 渗透(作用)
penetration of seawater 海水浸渍
penetration rate 渗透速率，渗透速度
percentage of moisture 含水率
perched ground water 悬着地下水，静滞地下水
perched water 悬着水
Perched water table 悬着地下水位，静滞地下水位
percolate 渗透，渗漏
percolating water 渗透水，渗漏水
percolation 渗透作用，渗漏作用
percolation rate 渗漏速率
percolation ratio 渗漏率
perennial drainage 常流水
permeability 渗透率，渗透性
permeable bed/stratum 透水层
phreatic decline 地下水位下降

phreatic fluctuation 地下水位升降，地下水位波动
phreatic hill 潜水灌丛沙堆
phreatic rise 地下水上升
phreatic solution 潜水溶解，潜水溶蚀
phreatic surface 地下水面，潜水面
phreatic water 潜水，地下水
phreatic zone 潜水层
physical weathering 物理风化
physico-mechanical means 物理化学固沙措施
physiological available water 生理有效水分
physiological drought 生理干旱
physiologically dry desert 生理干旱荒漠
phytogenic dune 植物沙丘(抛物线沙丘)
picket fence 防沙栅栏
pimple mound 小残丘
pit 水蚀穴；坑，井；深渊
Pitot tube 毕托管
pitting 蚀穴
plain of denudation 剥蚀平原
plain of desert leveling 荒漠侵蚀平原；干蚀平原
plane of deflation 风蚀面
plane of denudetion 剥蚀面
planorasion 风力坡蚀
plantation forest 人工林
plasmoeremion 沙漠
playa basin 干湖盆
pore pressure 孔隙压力
pore size 孔隙度，孔径大小
pore space 孔隙，孔隙空间
pore structure 孔隙构造
pore volume 孔隙容积
pore water 孔隙水
porewater pressure 孔隙水压力
porous fence 多孔栅栏，透风栅栏
potamic transport 水流搬运
proluvium 洪积物
protected area 防护区

protection action 防护作用
protection forest 防护林
protection of nature 自然保护
protective effect 防护效益
protective fence 防沙栅栏
protective forest belt 防护林带
psammophyte 沙生植物
pyramid dune 金字塔沙丘
pyramidal dune, polypyramid 金字塔沙丘，多边塔形沙丘(星状沙丘)

Q

qord 月牙形沙丘，金字塔沙丘
quantity of rainfall 降雨量
quantity of precipitation 降水量
quick sand, quicksand 流沙
quick-growing plantation 速生人工林

R

radial dune 辐射状沙丘(金字塔沙丘)
rain gush 暴雨
rain intensity 雨量强度
rain pit/pitting 雨滴坑，雨坑
rain shadow 雨影，雨影区
rain shadow effect 雨影效应
rain shower, rainshower 阵雨
rain splash 雨滴溅击侵蚀
rain-bearing wind 载雨风
raindrop 雨滴
raindrop detachment 雨滴分离作用，雨滴剥蚀作用
raindrop pit 雨滴坑
raindrop sand crater 雨滴沙坑
rainfall 降雨，降雨量
rainfall extreme 雨量极值
rainfall intensity 降雨强度
rainfall variability 雨量变率
raininess 雨量强度

rainmaking, rain-making 人工降雨
rainsplash erosion 雨水溅击侵蚀
rainsplash transport 雨水溅击搬运
rainy period/season 雨季
rainy year 丰水年
rain-wash, rainwash 雨水冲刷，雨水冲刷物
rainwash erosion 雨水冲刷侵蚀
rate of erosion 侵蚀速率
rate of fan deposition 冲积扇沉积速率
rate of infiltration 渗入速度，渗入率
rate of sand drift, rate of sand transport 输沙率
rate of sediment accumulation 泥沙淤积速率
rate of uplift 上升速率
ravine wind 峡谷风
RDD=resultant drift direction 合成输沙方向
RDP 合成输沙势
reactivation 再活化
recharge water 补给水，回灌水
reclamation 改造；土壤改良
reclamation of desert lands 荒漠土地改良
reclamation of desertified land 荒漠化土地治理改良
redeposit 再沉积
redeposited loess 次生黄土，再沉积黄土
redeposition 再沉积作用
redesertification 荒漠化逆转
reestablisment of vegetation cover 重建植被
reflective index, reflectivity 反射率
reforestation, forest regeneration 人工造林，森林恢复
regolith zone 风化层
regraded alkali soil 再生碱土
regraded saline soil 再生盐[渍]土
regression 退化，后退
rehabilitation of lands 土地恢复
relative altitude 相对高度
relative amount of sand drift/migration 相对输沙量
relative humidity 相对湿度

relative rate 相对速率
relative rate of sand transport 相对输沙率
relative sand drift 相对输沙量
relative sand-moving capability of wind 相对输沙率
relict dune 残遗沙丘
repeated deposition 反复沉积
repeated erosion 反复侵蚀
repose angle 休止角
reptating grain 蠕移颗粒
reptation 蠕移
reptation length 蠕移长度
reservoir lake 蓄水湖
residual hill 残丘
resultant wind 合成风
retarding evaporation 减速蒸发
retention ability 持水力
retention/precipitation ridge 滞留沙垄，沉积沙垄，积沙沙垄
return wate/flow 回流水，回流
revegetation 植被重建
reversed erosion 逆向侵蚀
reversed wind 反向风
reversing barchanoid ridge 反向新月形沙垄
reversing dune 反向沙丘
reversing ridge 反向沙垄
revolving storm 旋风
rework 再作用，再沉积
reworked sand 再沉积沙
ridge 沙垄，丘脊
rill erosion 细流侵蚀
ripple 沙纹
ripple deposit 沙纹沉积[物]
ripple mark 波痕
ripple movement 沙纹移动
ripple truncation surface 沙纹削蚀面
ripple wave length 沙纹波长
river deposit/deposition 河流沉积[物]
river erosion 河流侵蚀

river flow transfer 调水，河水调运
river head 河源
river junction 汇流点
river sand 河沙
river suspended drift 河流悬浮物质
river terrace 河流阶地
riverine desert 河岸荒漠
rounded cobble 圆砾(圆中砾)
rounded mound 圆形沙堆(穹状沙丘)
rounded sand 磨圆沙(粒)
rounding 磨圆[作用]
rouring 掏蚀
rubble flow 泥石流，碎石流
rubble land 石漠
rubby soil 砾质土
running sand 流沙
running water 流水
runoff coefficient 径流系数
runoff erosion 径流侵蚀
runoff infiltration 径流渗透[作用]
runoff intensity 径流强度
runoff-inducement, artificial inducement 集水，人工集水
runoff-producing storm 形成径流的暴雨

S

S by E=south by east 南偏东
S by W=south by west 南偏西
sabulous clay 沙质黏土
saline depression 盐化低地
saline desert 盐漠
saline lake 盐湖
saline playa, salt playa 干盐湖
saline seep, saline spring 盐泉
saline soil 盐渍土
saline solution 盐溶液，盐溶解
saline water conversion 咸水淡化
saline-alkali soil 盐碱土

salineness 含盐度
salinificaiton 盐化
salinity 盐度，盐渍度
salinity tolerance 耐盐性
salinization 盐化，盐渍化
salinization control 盐渍化治理，盐渍化防治
salinize 盐化
salinized soil 盐化土
salin-sodic soil 盐碱土
salt accumulation 积盐，盐分积累
salt bed 盐层
salt cone 盐锥
salt crust 盐结皮，盐结壳，盐壳
salt crust soil 盐壳土
salt damage 盐害
salt desert 盐漠
salt dome 盐穹，盐丘
salt endurance 耐盐性
salt resistance 抗盐性
salt tolerance 耐盐性，抗盐性
salt waste 盐漠
salt weathering 盐类风化[作用]
saltating 跃移
saltating fraction 跃移粒级
saltating grain, saltation particle/grain 跃移颗粒
saltating/saltation sand 跃移沙
saltation discharge 跃移输沙量
saltation length 跃移长度
saltation load 跃移质
saltation path 跃移轨迹
saltation path-length 跃移轨迹长度
saltation transport 跳跃搬运，跃移搬运
saltern 碱土
salty water 咸水，盐水
salty water soil 盐化土壤
sanctuary 禁猎区，保护地
sand accumulation 积沙
sand afforestation 沙地造林，沙荒造林

sand bank 沙坝，沙洲
sand bar 沙洲
sand bearing wind 输沙风，携沙风，风沙流
sand blast 吹沙磨蚀
sand blast action 风沙磨蚀作用
sand blasting 风沙磨蚀[作用]
sand body 沙体
sand carrying capacity 输沙量
sand clay 沙质黏土
sand clay loam 沙质黏壤土
sand deposit 沙沉积物
sand desert, sandy desert 沙漠
sand discharge 输沙量
sand dome 沙堆，沙穹丘
sand drift 风成沙堆积，沙堆，输沙
driving wind, sand-driving wind 挟沙风，风沙流
sand dune 沙丘
sand dunc fixation/stabilization 沙丘固定
sand fall 沙瀑布
sand feature 沙漠景观，风沙地貌
sand fence 防沙栅栏，沙障
sand fixation 固沙
sand flood 沙暴
sand flow, sandflow, sand flowage 沙流，风沙流
sand flow deposit 沙流沉积[作用]
sand flow rate 沙流速率，输沙率
sand flowage 沙流，风沙流
sand fraction 沙粒级
sand ground 沙地
sand haze 沙霾
sand hill 沙岗，沙山
sand hillock 沙堆，小沙岗
sand hole 沙坑
sand laden wind 携沙风，挟沙风，风沙流
sand mass 沙体
sand massif, sand mountain 沙山
sand mobility 沙的移动性
sand mound 沙堆

sand mounds topography 沙堆地形，沙堆地貌
sand movement by wind 风沙移动
sand moving capacity 输沙量
sand moving wind, sand-moving wind 输沙风
sand of storm 暴风沙
sand patch 沙斑，积沙地段
sand pillar 沙旋风
sand plain, sandy plain 沙原，沙质平原
sand ridge, sandridge 沙垄
sand rose 沙玫瑰(图)
sand ripple 沙纹
sand shadow 沙影堆积，背风积沙
sand sheet 沙片
sand sierra 沙山
sand size 沙粒级，沙粒径
sand soil 沙土
sand stabilization and reforestation 固沙造林
sand storm, sandstorm 沙暴
sand stringer, sand streak 沙带
sand supply 供沙[量]
sand thickness 沙层厚度
sand tonado 沙龙卷
sand tract 沙地
sand transport 输沙
sand transport rate 输沙速率
sand trap 沙陷阱，阻沙障碍物
sand trapping 沙陷阱作用
sand undulation 起伏沙丘
sand wave, sand-wave 沙波
sandblasting 风沙磨蚀作用
sand-break 防沙林，防沙障
sand-carrying capacity 输沙率，输沙量
sand-covered hillock 覆沙小丘
sand-drag 沙拖曳，沙辫
sand-dune stabilization 沙丘固定
sanded farmland 沙化耕地，风蚀性耕地
sanded soil 沙化土壤，风蚀性土壤
sand-fixing area 固沙区

sandflow 沙流，风沙流
sand-hazard area 沙害区域
sandification, sanding 沙化
sandkey 小沙洲
sand-moving capacity 输沙量
sand-moving power 输沙率
sand-protecting plantation 人工防沙林
sandridge desert 沙垄沙漠
sand-size aggregate 沙粒级团聚体
sand-size fraction 沙粒级
sand-size particle 沙粒级颗粒
sand-supply condition 供沙条件
sand-transmitting form 输沙方式
sandy beach-dune ridge 沙质海岸沙丘垄
sandy clay loam 沙质黏壤土
sandy desert 沙漠
sandy desertification 沙漠化
sandy hillock 沙堆
sandy loam 砂壤土
sandy silt loam 沙质粉砂壤土
sandy silt soil 沙质粉砂土
sandy soils 沙质土
sandy-pebble desert 沙砾质荒漠，沙砾质戈壁
saprolite 半风化体；风化层；残积层
saprolith 风化物；风化土
saturated humidity 饱和湿度
saturated sand flow 饱和沙流，饱和风沙流
saturation capacity 饱和量
saturation overland flow 饱和地表径流
scour 冲刷
scour and fill 边冲边淤
scour channel, scour trough 冲蚀槽
scrub vegetation 灌丛植被
SE by E 东南偏东
SE by S 东南偏南
sea level elevation 海拔高度
seasonal drought 季节性干旱
seasonal rainfall 季节性降雨

seasonal succession 季节性演替
second-order channel 二级水道
sedentary weathering zone 残积风化层
sedentary product 风化产物
sediment 沉积物
sediment accumulation 泥沙淤积[作用]
sediment concentration 含沙量
sediment discharge/load 输沙量
sediment sorting 颗粒分选[作用]
sediment supply 泥沙供应[量]
sediment transport 泥沙搬运；风成沙搬运
sediment transport rate 输沙率
sedimentary basin 沉积盆地
sedimentary strata 沉积层
sedimentary supply 沉积供应
sedimentation 沉积作用
sediment-moving desert wind 荒漠输沙风
seepage 渗出，渗漏
seepage discharge 渗流量
seepage flow 渗流
segment 碎片；新月形沙丘段
seif, sif, sief, sayf((复数 siouf)赛夫沙丘
seif dune chain 赛夫沙丘链
seifdune, seif-dune 赛夫沙丘
semi-fixed dune 半固定沙丘
semi-frutex 半灌木
semi-immobile dune 半流动沙丘
semilunar dune 新月形沙丘
semi-mobile sand 半流动沙地
semi-stabilized dune 半固定沙丘
shadow dune 风影沙丘
shear sorting 剪切分选[作用]
sheet erosion 片蚀
sheet flood, sheet-flood 片流，漫洪
sheet wash, sheetwash 片蚀
sheetflood erosion 片流侵蚀
shelter belt, sheltbelt, protective belt 防护带，防护林带

shelter forest 防护林
sheltering effect 掩蔽作用，掩蔽效益
shower, shower rain, showery rain 阵雨
shrub cover density 灌丛植被密度
shrub grassland 灌丛草原，灌丛草地
shrub-coppice dune 灌丛沙丘
shrub-topped mound 灌丛土丘，灌丛沙堆
sigmoidal dune S 形沙丘
silk road 丝绸之路
silt 粉沙
silt clay 粉沙黏土
silt content 粉粒含量
silt fraction, silty fraction 粉沙粒级
silt loam 粉砂壤土
silt-size particle 黏粒
silty stony desert 粉沙质石漠
silvicultural reclamation 造林改良土壤；农林改良土壤
simple crescentic dune 简单新月形沙丘
simple crescetic dune ridge 简单新月形沙丘垄
simple dome shaped dune 简单穹状沙丘
simple dominant wind 单向盛行风
simple linear dune 简单线形沙丘
simple mound 简单沙丘，简单沙堆（新月形沙丘）
simple parabolic dune 简单抛物线沙丘
simple star dune 简单星状沙丘
simple streak 简单沙带
singing sand, acoustical, barking, booming, musical, roaring, sonorous, sounding, whispering, whistling sand 鸣沙，响沙，轰鸣沙
size grading 粒度级配
size parameter 粒度参数，粒级参数
size sporting process 粒级分选过程
slacktip 崩坍
sliding 滑动
slope angle, angle of slope 坡角

slope aspect, aspect of slope 坡向
slope base 坡脚
slope deposit 坡积物
slope runoff 坡地径流
slope transport 坡地搬运
slope wash 坡积物
snowmelt runoff 融雪径流
soil aeration 土壤通气性
soil aggregate 土壤团聚体
soil and water conservation 水土保持
soil and water loss 水土流失
soil assessment 土壤评价
soil builder 保土植物
soil capillarity 土壤毛管性
soil consistence, soil consistency 土壤结持度
soil constituent 土壤成分
soil corrosion 土壤溶蚀[作用]
soil cover 土被，土壤
soil crust 土壤结皮，土壤结壳
soil degradation 土壤退化[作用]
soil denudation 土壤侵蚀[作用]
soil depletion, soil exhaustion 土壤耗竭
soil deterioration 土壤破坏[作用]
soil development 土壤发育
soil drought 土壤干旱
soil erodbility 土壤可蚀性
soil erosion 土壤侵蚀
soil/solifluction flow 泥流
soil forming process 成土过程
soil improvement 土壤改良
soil mechanical composition 土壤机械组成
soil microorganism 土壤微生物
soil moisture 土壤水分
soil moisture conservation 土壤保墒
soil moisture content 土壤含水量
soil moisture retention 土壤持水性，土壤保水性
soil particle size 土壤粒度，土壤粒径
soil permeability 土壤渗透性

soil plasma 土壤细粒物质
soil profile differentation 土壤剖面分异[作用]
soil reclamation 土壤改良
soil reservoir 土壤蓄水层
soil salinity 土壤盐度
soil texture 土壤质地
soil water balance 土壤水分平衡
soil zonality 土壤地带性
soil-protecting plant 保土植物
soil-water storage 土壤贮水量
solitary barchan 孤立新月形沙丘
solodic soil 脱碱化土壤
solonchak desert 盐土荒漠
solonetz soil 碱土
solonetzification 碱化作用
soluble salt 可溶性盐
solum 风化层（土壤剖面顶层）
solute 溶质，溶解物
solute transfer 溶质移动
solution basin 溶蚀盆地
solution etching 溶蚀
solution hollow 溶蚀凹地
solution pit 溶蚀坑
solutional attack 溶蚀作用
solutional karstic phenomenon 溶蚀喀斯特现象
sorting 分选[作用]
sorting coefficient/index 分选系数
soueaster, southeaster 强东南风，东南大风
source of dust 尘源
south by east 南偏东
south by west 南偏西
south pole 南极
southeast (SE) 东南
south-east by east 东南偏东
southeast by south 东南偏南
south-east trade wind, southeast trades 东南信风
Southern Hemisphere 南半球
southern latitude 南纬

southward 向南
southwest (SW) 西南
south-west by south 西南偏南
south-west by west 西南偏西
southwest monsoon 西南季风
sparse vegetation 稀疏植被
spatial zonation 空间成带现象，空间带状分布
sprinkler/squirting irrigation 喷灌
stabilited wind-erosion feature 固定风蚀地形
stabilization 稳定[作用]；固定[作用]
stabilization of shifting sand 流沙固定
stabilize free dune 固定活动沙丘
stabilized dune 固定沙丘
stabilized longitudinal dune 固定纵向沙丘
stabilized mega-dune 固定大沙丘
stabilized sands 固定沙地
stagnamt water 停滞水
star dune in chain 星状沙丘链
star-like dune 星状沙丘
stellate dune, star dune 星状沙丘
steppe meadows 草原草甸
steppification, steppization 草原化
stone desert, stone mosaic, stone pavement 石漠
stone mantle, stony desert 石漠，砾漠
stoss side 迎风面，迎风侧
stoss slope, windward slope 迎风坡
stream transportation 河流搬运[作用]
streamflow 河川径流
streamline dune 流线型沙丘
stripping 剥蚀[作用]
strong breeze 强风(六级风)
strong gale 烈风(九级风)
subsidiary arm 副臂，子臂
subsidiary crescentic dune 副新月形沙丘，子新月形沙丘
subsidiary ridge 副沙垄，子沙垄
summit 山顶，丘顶
surface creep 表面蠕移，表面蠕移运动

surface crust 地面结皮
surface drainage 地表排水，明沟排水；地表水系
surface particle concentration 地面粗粒化
surface reactivation 地面再活化
surface roughness 表面粗糙度，地面粗糙度
surface runoff 地表径流
surface soil 表土
suspended matter 悬浮物质
suspended particles 悬移颗粒
suspended sediment concentration 悬浮质含量
suspended water 悬着水
suspension 悬移，悬移运动
suspension current 悬浮流
suspension load 悬移质
SW = southwest 西南
SW by S 西南偏南
SW by W 西南偏西
sweat 凝结水
sweet water 淡水
symmetrical ripples 不对称沙纹

T

table sands 固定沙地
takoit 灌丛沙堆
texture grade 土壤质地等级
thermal turbulence 热湍流
thermal wind 热成风
thermos-isopleth 等温线
threshold 临界值
threshold drag velocity 起动摩阻速度
threshold friction velocity 起动摩擦速度，起动摩擦风速
threshold of entrainment/movement 起动风速
threshold shear velocity 起动切应风速
threshold speed/velocity 起动速度，起沙风速
threshold velocity for sand movement 起沙风速
through flow, throughflow 渗[透]流
time of low water 枯水期

toe 迎风坡脚
tongue hill 舌状丘
total deposition 总沉积量
total discharge 总输沙量；总流量
total pore space 总孔隙
total porosity 总孔隙度
total transport rate 总输沙率
touradon 沙丘
transition zone, transitional belt 过渡带
transpiration 蒸腾[作用]
transport rate 搬运速率，输沙率
transportation capacity 搬运量
transported road 搬运质
transported material 搬运物质
transverse bar 横向沙洲
transverse crescentic link 横向新月形沙丘链
transverse draa 横向沙丘链(垄)
transverse dune 横向沙丘
transverse mega-dune 横向巨型沙丘
transverse ridge 横向沙垄
transverse ripple 横向沙纹
transverse sand wave 横向沙波
transverse type dune 横向类型沙丘
travelling dune 流动沙丘
turbulence, turbulence current/flow 湍流

U

uncropped location/soil 荒地
under grazing 轻度放牧
undercut 底切，暗掘，掏蚀
underflow 地下水流，潜流
underground reservoir 地下水库
underground residue 地下水贮量
underground waters 地下水
underground watering 浸润灌溉
underload 低含沙量
undershrub 半灌木，小灌木
unfixed sand 流沙

unidirectional dune field 单向沙丘区
unidirectional flow 单向流
uni-directional wind 单向风
unimodal distribution 单峰分布
unstabilized transverse dune 不固定横向沙丘
unstable channel 不稳定河床
upstream 上游
unulation 波状起伏，波状沙丘
updraft 向上气流
upland water 地表水
upward-sloping portion 迎风坡部位
upwind direction/side 上风向，迎风侧
upwind slope 迎风坡
U-shaped dune U 形沙丘，即抛物线沙丘
U-shaped parabolic dune U 形抛物线沙丘

V

vadose 渗流
vadose solution 渗溶
variable wind direction 多变风向
variable wind regime 多变风况
variability 变率
vegetal dune stabilization 植物沙丘固定，植物固沙
vegetation cover 植被；植被盖度
vegetation restoration 植被恢复
vegetation wind-break 植被防风带
vertical erosion 垂直侵蚀，下切侵蚀
very coarse sand 极粗沙
very fine sand 极细沙
very fine sandstone 极细[粒]砂岩
very fine sandy loam 极细砂壤土
very poor drained 排水不良，排水极差
violent storm/wind 暴风(十一级风)
volume weight 容重
V-shape dune V 形沙丘(抛物线沙丘)

W

W = west 西

W by N 西偏北
W by S 西偏南
wandering dune 流动沙丘
warp 淤积物，沉积物；弯曲，曲折
wash 冲积物，冲刷，侵蚀；干河床
wash erosion 冲蚀
wash load 冲刷搬运物（泥沙）
wash-controlled surface 冲刷侵蚀
waste mantle 风化残积层
water absorbing capacity 吸水量
water and erosion control structure 水土保持工程
water borne sand grain 水成沙粒
water conservation 水分保持
water consumption 耗水[量]
water deficit 水分亏缺
water delivery 供水
water demand/requirement 需水量
water erosion 水蚀
water of infiltration 渗入水，入渗水
water of saturation 饱和水
water penetration 水分渗透[作用]
water permeability 透水性
water retaining capacity 持水量
water retention property 保水性
water salinity 水盐渍度
water storage capacity 贮水量
water supply 供水，给水
water utilization 水利用
water-control 治水，水利措施
water-deposited sediment 水成沉积物
water flood 洪水
water-holding pore 持水孔隙
watering 灌水，浇水，给水，引水
water-laid deposit 水流沉积[物]
watershed 分水岭流域，集水区
water-storage capacity 贮水量
water-tight layer 不透水层
wear away 磨损

weathered layer 风化层
weathered mantle 风化覆盖物，风化覆盖层
weathering 风化作用
weathering intensity 风化强度
weathering landform 风化地貌
weathering pit 风化穴
weathering residues 风化残积物
weather-side 上风的，迎风的
well irrigation 井灌
west by south 南偏西
west longitude 西经
west side 西坡
westerlies, westerly belt 西风带
wet year 丰水年，湿润年
wetland 湿地
whole gale 狂风（十级风）
wide bimodal wind regime 宽双峰风况
wide unimodal wind 宽单峰风
wild flooding irrigation 滥灌，漫灌
wilting coefficient 凋萎系数
wilting percentage 凋萎含水率
wind base 迎风坡脚
wind/wind-break belt 防风林带
wind blown sediment, wind deposit 风积物
wind corrosion 风蚀
wind deposited sand, wind drift 风积沙
wind direction 风向
wind directionality 风向性
wind desiccation 风干作用
wind drift sand 飞沙，风沙，风沙流
wind drying soil 风干土
wind erodibity 风蚀性
wind erosion furrow 风蚀洼地
wind facetted pebble 风棱石
wind force, wind strength 风力
wind hodograph, hodograph 风向图
wind hole 风蚀穴
wind profile 风速廓线

wind resistance 抗风性
wind rifting, wind drifting 风蚀作用
wind ripple 风成沙纹
wind rose 风玫瑰(图)，风向图
wind scouring 风蚀
wind speed, wind velocity 风速
wind transport 风搬运
wind tunnel 风洞
wind vector 风矢量
wind velocity profile 风速廓线
wind worn pebbie 风蚀砾石
windbelt 防风林带
wind-blown naterial 风吹物质；风蚀物质
windhlown sand 风成沙，风积沙，
windblown silt 风尘，风积尘
windbreak 防风屏障，防风林带
wind-break belt 防风林带
winddrift dune, windrift dune 风积沙丘
wind-driven current 风沙流
wind-eroded feature 风蚀景观，风蚀地形
wind-eroded hill 风蚀丘陵，风蚀垄岗
wind-eroded hollow 风蚀凹地
wind-formed ripple mark 风成波痕，风成沙纹
wind-grooved boulder 风蚀砾
winding groove 风蚀沟
wind-moulded landscape 风蚀景观，风蚀地形
wind-protection plantation 防风林
wind-sorted material 风选物质
windward, windward/side 向风面，迎风面

windward/stoss slope 向风坡，迎风坡
wind-wise length 沙丘长度
windworn pebble, wind worn pebble 风蚀砾石
wing-to-wing width 沙丘宽度
winter monsoon 冬季季风
winter rain 冬雨
winter resistance 抗寒性，耐寒性

X

xerochore 无水荒漠区，干旱荒漠
xeromorphic soil 旱成土
xeromorphic vegetation 旱生植被

Y

yardang, yarding, jardang 风蚀土墩，风蚀土脊，龙堆，白龙堆，雅丹
yardang-like wind erosion 雅丹状风蚀
year with abundance of water 丰水年
year with low water 枯水年

Z

zonality 地带性
zonation 成带现象
zone of accumulation 淀积层
zone of aeration 包气层
zone of capilary saturation 毛管水饱和层
zone of leaching 淋溶层
zone of trade wind 信风带